An Introduction to the Rasch Model with Examples in R

Chapman & Hall/CRC
Statistics in the Social and Behavioral Sciences Series

Series Editors
Jeff Gill, Steven Heeringa, Wim J. van der Linden, Tom Snijders

Recently Published Titles

Big Data and Social Science: Data Science Methods and Tools for Research and Practice, Second Edition
Ian Foster, Rayid Ghani, Ron S. Jarmin, Frauke Kreuter and Julia Lane

Understanding Elections through Statistics: Polling, Prediction, and Testing
Ole J. Forsberg

Analyzing Spatial Models of Choice and Judgment, Second Edition
David A. Armstrong II, Ryan Bakker, Royce Carroll, Christopher Hare, Keith T. Poole and Howard Rosenthal

Introduction to R for Social Scientists: A Tidy Programming Approach
Ryan Kennedy and Philip Waggoner

Linear Regression Models: Applications in R
John P. Hoffman

Mixed-Mode Surveys: Design and Analysis
Jan van den Brakel, Bart Buelens, Madelon Cremers, Annemieke Luiten, Vivian Meertens, Barry Schouten and Rachel Vis-Visschers

Applied Regularization Methods for the Social Sciences
Holmes Finch

An Introduction to the Rasch Model with Examples in R
Rudolf Debelak, Carolin Stobl and Matthew D. Zeigenfuse

Regression Analysis in R: A Comprehensive View for the Social Sciences
Jocelyn H. Bolin

For more information about this series, please visit: https://www.routledge.com/Chapman--HallCRC-Statistics-in-the-Social-and-Behavioral-Sciences/book-series/CHSTSOBESCI

An Introduction to the Rasch Model with Examples in R

Rudolf Debelak, Carolin Stobl and
Matthew D. Zeigenfuse

CRC Press
Taylor & Francis Group
Boca Raton London New York

CRC Press is an imprint of the
Taylor & Francis Group, an **informa** business
A CHAPMAN & HALL BOOK

First edition published 2022
by CRC Press
6000 Broken Sound Parkway NW, Suite 300, Boca Raton, FL 33487-2742

and by CRC Press
4 Park Square, Milton Park, Abingdon, Oxon, OX14 4RN

CRC Press is an imprint of Taylor & Francis Group, LLC

Library of Congress Cataloging-in-Publication Data

Names: Debelak, Rudolf, author. | Strobl, Carolin, author. | Zeigenfuse, Matthew D., author.
Title: An introduction to the Rasch model with examples in R / Rudolf Debelak, Carolin Strobl and Matthew D. Zeigenfuse.
Description: First edition. | Boca Raton : CRC Press, 2022. | Series: Chapman & Hall/CRC statistics in the social & behavioral sciences | Includes bibliographical references and index.
Identifiers: LCCN 2021056957 (print) | LCCN 2021056958 (ebook) | ISBN 9781032265582 (hardback) | ISBN 9781138710467 (paperback) | ISBN 9781315200620 (ebook)
Subjects: LCSH: Rasch models. | Psychology--Statistical methods. | Educational tests and measurements. | R (Computer program language) | Psychometrics--Data processing.
Classification: LCC BF39.2.I84 D43 2022 (print) | LCC BF39.2.I84 (ebook) | DDC 150.28/7--dc23/eng/20220318
LC record available at https://lccn.loc.gov/2021056957
LC ebook record available at https://lccn.loc.gov/2021056958

ISBN: 978-1-032-26558-2 (hbk)
ISBN: 978-1-138-71046-7 (pbk)
ISBN: 978-1-315-20062-0 (ebk)

DOI: 10.1201/9781315200620

Typeset in CMR10 font
by KnowledgeWorks Global Ltd.

Publisher's note: This book has been prepared from camera-ready copy provided by the authors.

R.D.: for Markus

C.S.: for Tonia, Konstantin, and Torsten

M.D.Z.: for Angela

Contents

Preface

The goal of this book is to provide an accessible introduction to the Rasch model (Rasch, 1960), an important tool in educational and psychological measurement named after the Danish mathematician Georg Rasch. The Rasch model mathematically describes the responses of a sample of test takers on a set of test items. The application of a mathematical model allows a precise description of the characteristics of test takers and test items. This in turn allows, for instance, the estimation of a person's abilities based on their test performance and the detection of flawed test items, which can affect the fairness of a test.

One hurdle for many attempting to learn about this model is the mathematical and statistical background required to do so. We have collected all of this background knowledge into the chapters and appendices of this book, with the hope of making the mathematics and statistics behind the Rasch model accessible to everyone. For this reason, this book is well-suited for use in an introductory class within a psychology or educational science program, or for learning on your own.

In many places in the book, you will find that we provide more detailed accounts of the steps involved in a computation than most other textbooks on this topic. We wanted to create a book that is accessible for a variety of audiences, and we have included this material to help readers with less mathematics training. Some readers may prefer focusing on the text and returning to the math sections later, if necessary.

In addition to the theory, we also provide a number of hands-on examples in R, which is a free, open-source software platform for statistics. We do not assume any prior experience with R and provide all of the information you will need to get started in this book. That said, this is not a general introduction to R. If you would like to use R for other applications, we recommend an introductory R textbook.

Acknowledgment

We thank you for choosing this book, and we would also like to thank a number of people for their help in preparing it, including Corinne Bircher for an early translation of portions of this book from German; Angela Zeigenfuse for language editing; Alina Gasser and Vinh Tong Ngo for copy editing later drafts; Shashi Kumar and Emre Akbulut for technical assistance; Julia Kopf, Basil Komboz, Dries Debeer, Lale Khorramdel, Chung Man (Mandy) Fong, Carolina Fellinghauer, Mirka Henninger, Urs Grob, Achim Zeileis, and Felix Zimmer for valuable hints and feedback on different parts and versions of the text and R code; Phil Chalmers, Alexander Robitzsch, Patrick Mair, and the late Reinhold Hatzinger for answering inquiries about their respective R packages; and our previous and current students (as well as our friends whom we used to tutor back when we were still students ourselves), because teaching them has helped us realize which parts of the theory and application need extra explanation. All the remaining errors and omissions are our own. We are grateful for any feedback.

Part I

Theory

1

Introduction

CONTENTS

Before introducing the Rasch model in the next chapter, we would briefly like to discuss the role of psychological and educational tests in different areas of research and practice and to compare and contrast the Rasch model with related measurement frameworks. Then, we provide an overview of the contents of this book.

1.1 The Role of Psychological and Educational Tests

Psychological and educational tests play an important role in many areas of research and practice. Tests have been developed for measuring a range of phenomena, from mathematical ability to job suitability to depression susceptibility. They are necessary because mental attributes such as mathematical ability and depression susceptibility are latent, meaning that their value cannot be directly observed. The process of using a test to learn about a mental attribute is known as psychological measurement.

Before a test can be used, it must be constructed. Typically, this entails the generation of a set of potential items by content specialists. For example, a test of high school algebra might contain items covering the quadratic formula and completing the square. These items are then critically evaluated to ensure that they satisfy different criteria. Items that do not meet these criteria need to be revised or excluded.

One common criterion is fairness (cf. American Educational Research Association, American Psychological Association, & National Council on Measurement in Education, 2014). For example, word problems are often used to test mastery of high school algebra. Because word problems involve reading, test takers' responses will be influenced by both their algebra and reading skills. This can put nonnative speakers at a disadvantage, thereby making them appear to have been less successful at mastering algebra when they were actually only less successful at understanding the wording of the algebra test items. While one could argue that reading skills are also important, and should indeed influence the results, e.g., for a test measuring college readiness, for a test that aims to measure only algebra proficiency such items can be considered as unfair.

Thus, it is important that everyone who uses psychological or educational tests understands how they are evaluated. Even if you are not developing tests yourself, you still need to be able to understand reports on test evaluation in order to choose the test that does the best job of measuring whatever you are interested in. Using a test that has not been rigorously evaluated can lead to unfair comparisons between people and can have serious consequences for their educational and professional advancement or for the treatment of their psychological symptoms.

The Rasch model and related models provide a powerful framework for evaluating tests and form an integral part of psychometric practice. They have been used to develop a number of widely known standardized achievement tests, including the GRE exam for graduate school admissions in the US, as well as large scale educational assessments, such as PISA and TIMMS. They have also been used to (re-)evaluate a variety of established tests in different areas of psychology, medicine and related disciplines.

1.2 The Rasch Model and Item Response Theory

When reading about the Rasch model, you may realize that it is often linked to or contrasted with other methodologies. Here, we will briefly list and attempt to "sort through" some of the concepts that are relevant to placing the Rasch model both historically and philosophically and refer the interested reader to more extensive presentations.

The Rasch model is sometimes treated as a representative of modern test theory to contrast it with classical test theory (CTT), which was the predominant approach in psychological test construction and evaluation for a long time. Important differences between the Rasch model and CTT are that the assumptions of the Rasch model are empirically testable (as you will see in detail throughout this book), and that it abolishes the "systematic confounding of what we measure with the test items used to measure it" (van der Linden, 2016–2018, in his Preface) that is inherent in CTT. A systematic comparison can be found, e.g., in the book of Andrich and Marais (2019).

Another framework for psychological test construction that the Rasch model relates to is item response theory (IRT). You might find the literature regarding the nature of this relationship to be somewhat contradictory. While some authors consider the Rasch model as part of IRT, others explicitly contrast it with IRT. The reason for this apparent discrepancy is that while the Rasch model and those models that are unequivocally labeled as IRT models have a strong mathematical relation and were developed in parallel historically, philosophically they stem from very different traditions (or paradigms; Andrich, 2004; Andrich & Marais, 2019).

For example, we will explain in detail in this book that one important property or assumption of the Rasch model is that all items have the same discrimination or slope. This is the basis for the special measurement properties of this model. However, items from a real test may not satisfy this strict assumption.

Models from IRT relax this assumption and allow items to have different slopes. These more flexible models allow us to describe behavior on a wider array of tests. Unfortunately, they do so at the expense of desirable measurement properties. Thus, we are left with two conflicting points of view: what we will call in the following the modeling point of view, which views IRT as a collection of statistical models used to "describe the items", and what we will call the measurement point of view, which views the Rasch model "as a means of quality control of the items" (citations from Bond & Fox, 2007).

Introductions to the Rasch model from the modeling point of view (but without the authors necessarily using this label) can be found in Embretson and Reise (2000) and Reckase (2009). The former work also provides an overview of the history of IRT, allowing the reader to place the most influential figures in the early development of IRT, while the latter extends IRT to multidimensional latent traits. These introductions cast the Rasch model as a special case of more complex IRT models, obtained by fixing one or more of the item parameters to a constant value (see Section 4.5.1). This varying degree of flexibility is visible in the shapes of the item response functions, which we will later display for the different models.

An even more flexible approach is nonparametric IRT, where the form of the item response functions is even less restricted. Instead of providing a specific form of the item response functions, nonparametric IRT models typically focus on meeting certain assumptions, such as monotonicity or unidimensionality, to allow measurements on an ordinal scale. Introductions to nonparametric IRT are provided, e.g., in van der Linden and Hambleton (1997), by Sijtsma (1998) and Sijtsma and van der Ark (2017). In this book, however, we will be focusing on parametric models.

Introductions to the Rasch model taking a measurement point of view are given in Bond and Fox (2007) and Andrich and Marais (2019). Both also include historical reviews of other important measurement traditions or concepts such as CTT and the Guttman Structure (Andrich & Marais, 2019) or conjoint measurement (Bond & Fox, 2007). The latter, in their preface, also cite a quote from (we assume Mark) Wilson that very humorously captures the modeling vs. measurement controversy: "[O]ne person's oversimplification is another person's strong measurement philosophy."

In addition to authors who clearly contrast their viewpoint as either measurement ("team Rasch") or modeling ("team IRT"), there are also authors who highly value the special characteristics of the Rasch model but are nevertheless willing to move on to a model with more parameters when it cannot account for the data from a given test. While these authors are strongly influenced by the measurement point of view, they explicitly use the label IRT as an umbrella term for all item response models including the Rasch model. For example, van der Linden and Hambleton (1997) state that "[f]ormally, the [Rasch] model is a special case of the Birnbaum model [...]. However, because it has unique properties among all known IRT models, it deserves a separate introduction" (we find the same view apparent in van der Linden, 2016–2018).

We realized when writing this book that we were socialized in this latter tradition and thus might in some places use IRT as an umbrella term that includes the Rasch model. We will also try our best to explain arguments and approaches from both the measurement and the modeling point of view. We hope that – while we probably cannot fully avoid stepping on either side's toes from time to time – through this approach, we provide enough information about the beauty and strictness of the Rasch model to enable the reader to appreciate each viewpoint.

One last comment on terminology: in this book we use the term Rasch model exclusively for the model for dichotomous items (with true/false or agree/disagree responses), and large parts of this book will concentrate on this model. There are also models for polytomous items (Partial Credit and Rating Scale models) that belong to the family of Rasch models in the sense that they assume equal slopes for all items. These models are introduced in Chapter 11.

1.3 Where You Will Find What in This Book

The first part of this book introduces the theory. We will start with a thorough introduction to the Rasch model in Chapter 2. We then introduce methods for estimating the parameters of the Rasch model in Chapter 3, and present a variety of approaches for empirically evaluating tests in Chapter 4.

The second part of the book is dedicated to the practical application in R (R Core Team, 2021). We have elected to use R, because it is free and provides all of the functionality necessary to perform standard item response theory analyses. Additionally, R can be extended to perform nonstandard analyses when necessary. In Chapter 5, we provide a brief introduction to the R language. In Chapter 6 through Chapter 9 we demonstrate for different R packages how to apply the theory from the first part of the book using real-world data.

The third part of the books gives an outlook beyond the dichotomous Rasch model. In this part we touch upon a variety of topics, which are not strictly part of an introduction to the Rasch model. Therefore, these topics are treated in less detail, but provide you with a "bigger picture" on related models and important application areas. This part of the book outlines, for example, models for polytomous items, the formulation of the Rasch model as a mixed model and computerized adaptive testing (CAT).

At the end of the book you will find mathematical and statistical background information in the form of two appendices. You will be referred to the appendices in those theory chapters that require their contents. You can then either go through the appendix systematically before starting the chapter in case you have not previously covered the contents, or you can go back later in case you only want to brush up on something. The last appendix provides solutions to the end-of-chapter exercises, including R commands for the exercises in the application chapters.

2

The Rasch Model

CONTENTS

The Rasch model is based on the intuition that a person's test responses depend on the value of a latent trait for that test taker. For example, we expect that a student who is good at mathematics will score higher on a mathematics test than a student who is not. In this way, the observable test responses can provide information about the unobservable latent trait value. If each student in a class completes a test, we expect that the student with the highest score has the highest ability, the one with the second highest score has the second highest ability, etc. Without this expectation, it would be impossible to justify giving better grades to students who earn higher test scores.

The Rasch model formulates this idea using a mathematical model. According to the model, every person has a probability of correctly answering every test item. This probability is determined by the ability of the person and the difficulty of the test item. Formulating the model in this way allows us to measure, or infer, ability from test scores.

The remainder of this chapter presents the Rasch model. We begin by describing the data matrix in Section 2.1. In Section 2.2, we introduce the Rasch model and intuitively explain why it makes sense as a model for test-taking behavior. Alternative representations of the Rasch model that you might find in other textbooks are covered in Section 2.3. We then present the assumptions implicit in the Rasch model in Section 2.4. These assumptions are important, because they determine the situations in which the Rasch model can be used for measurement. We should not apply the Rasch model when its assumptions do not seem reasonable or are not supported by data. If we find this to be the case, depending on our view we either have to exclude or improve items that are not in line with the Rasch model, or use a more flexible model.

Please note that this chapter is intentionally thorough. It may contain more formulas and intermediate steps than are needed to understand the basics of the different procedures and apply them to data. We have included this information because other textbooks may not provide enough detail for those readers who do want to follow specific computations. However, you are free to skip some of the formulas and computations and concentrate on the surrounding text – especially if you are reading about the Rasch model for the first time.

2.1 The Data Matrix

	Item					
Person	1	2	3	4	5	6
1	0	1	0	1	0	1
2	0	1	1	0	1	1
3	0	1	1	1	0	0
4	1	0	0	1	0	0

TABLE 2.1: Data matrix for a six-item test that has been completed by four people.

The Rasch model uses a person's pattern of correct and incorrect responses to measure ability or attitude. When dealing with multiple people, these responses are collected into a simple table or matrix. Each correct answer or agreement will be listed as 1 and each wrong answer or disagreement will be listed as 0. For example, suppose that Table 2.1 contains the data from four people taking a six-item mathematics test. It shows that Person 1 answered items 2, 4, and 6 correctly, Person 2 answered items 2, 3, 5, and 6 correctly, etc.

	Item						
Person	1	2	...	i	...	$I-1$	I
1	u_{11}	u_{12}		u_{1i}		$u_{1,I-1}$	u_{1I}
2	u_{21}	u_{22}		u_{2i}		$u_{2,I-1}$	u_{2I}
$\vdots$							
p	u_{p1}	u_{p2}		u_{pi}		$u_{p,I-1}$	u_{pI}
$\vdots$							
$P-1$	$u_{P-1,1}$	$u_{P-1,2}$		$u_{P-1,i}$		$u_{P-1,I-1}$	$u_{P-1,I}$
P	u_{P1}	u_{P2}		u_{Pi}		$u_{P,I-1}$	u_{PI}

TABLE 2.2: Data matrix $\boldsymbol{u}$ for the general case in which an I-item test is completed by P people.

Table 2.2 shows the general case where P people complete a test with I items. In this case, the data matrix, here denoted $\boldsymbol{u}$, contains P rows and I columns. Each row $\boldsymbol{u}_{p\cdot}$ of the matrix gives the responses of a single person p to all of the test items. Each column $\boldsymbol{u}_{\cdot i}$ gives all of the observed responses to a single test item i. The response of person p to item i is u_{pi}, where each u_{pi} is a placeholder for whatever values we observe. The data matrix from an actual test will have the u_{pi} filled in with zeros and ones.

The Rasch model predicts the probability that each person correctly responds to each test item given the person's ability. A probability is a number between zero and one expressing certainty over whether some event, like a correct response to a test item, will occur. A probability of one indicates that an event is guaranteed to occur, while a probability of zero indicates that it is guaranteed not to occur. A probability of $1/2$ indicates that the event is as likely to occur as not.

It makes sense to use probability, because test taking behavior is complicated and depends on a variety of factors other than ability. For example, a person who slept poorly the night before the test might perform worse than expected according to their ability. Alternatively, a person might simply get lucky and correctly guess a few items whose answers they didn't know. Probabilities reflect the uncertainty inherent in predicting test performance solely from ability. The Rasch model attributes occurrences like lack of sleep up to chance in order to avoid modeling them directly.

We will need to introduce a small amount of notation in order to effectively work with probabilities. First, we use $\Pr(E)$ to denote the probability of some event E. For example, E could be the event that it rains tomorrow, in which case, $\Pr(E)$ is the probability that it rains tomorrow. Second, we use random variables U_{pi} to represent responses. Intuitively, we might think of each U_{pi} as a placeholder for person p's actual response to item i, u_{pi}, before the test is administered. We combine these two pieces of notation into probability statements.

2.2 The Item Response Function

The Rasch model formalizes two common sense observations about test-taking behavior. The first one is that people with higher ability tend to answer more test items correctly, regardless of what items are chosen for the test. The Rasch model formalizes this observation by asserting that the probability of correctly answering an item increases with increasing ability. The second observation is that test items with a higher difficulty tend to be answered correctly less often. The Rasch model formalizes this by asserting that the probability of correctly answering an item decreases with increasing difficulty.

We define θ_p to be the ability of person p and β_i to be the difficulty of item i. For the remainder of this chapter, we'll assume that θ_p and β_i are known. This might seem a little strange, since the reason we administered the test is to learn each person's ability. However, by writing the model as if we already know θ_p and β_i, we can use statistical inference to "invert" the model to learn their values. This process, which is called parameter estimation, will be presented in the next chapter.

For known values of θ_p and β_i, the Rasch model predicts that the probability that person p correctly answers item i is

$$\Pr(U_{pi} = 1 \mid \theta_p, \beta_i) = \frac{\exp(\theta_p - \beta_i)}{1 + \exp(\theta_p - \beta_i)}. \tag{2.1}$$

The vertical bar in $\Pr(U_{pi} = 1 \mid \theta_p, \beta_i)$ indicates that the expression denotes a conditional probability. The placement of θ_p and β_i to the right of that bar indicates that they are known. It is typically read as "the conditional probability that $U_{pi} = 1$ given θ_p and β_i". The expression $\exp(\cdot)$ represents the exponential function $\exp(x) = e^x$. For consistency, we will use the $\exp(x)$ form throughout, because it is more readable for complicated exponents.

2.2.1 Ability and Difficulty

Figure 2.1 illustrates the relationship between ability and the probability of a correct response that is predicted by the Rasch model for a given difficulty. Each of the three curves is known as an item characteristic curve (ICC) and shows the probability of a correct response $\Pr(U_{pi} = 1 \mid \theta_p, \beta_i)$ that is predicted by the Rasch model for a given difficulty β_i at each level of ability θ_p. We can read off the predicted probability for a particular ability by

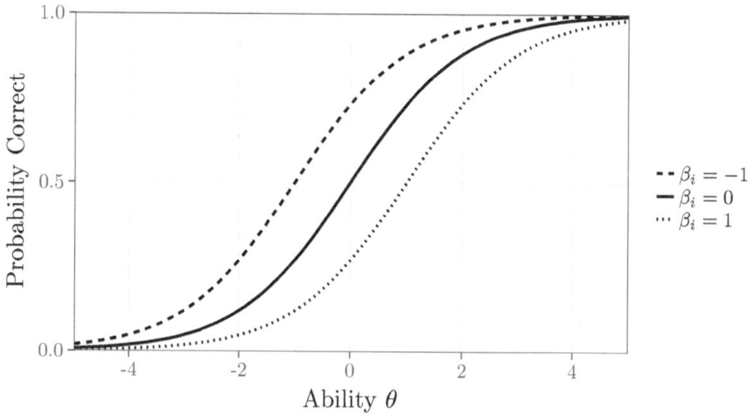

FIGURE 2.1: Rasch ICCs for different difficulty parameters. The x-axis corresponds to ability and the y-axis to the probability of a correct response.

tracing a vertical line from the desired ability to the ICC and then a horizontal line to the y-axis. The value that you hit is the probability of a correct response (often abbreviated to probability correct in the following). For example, if we wanted to find the probability that a person whose ability $\theta_p = 0$ correctly responds to an item whose difficulty $\beta_i = 0$, we could trace a vertical line starting at the value 0 on the x-axis to the solid ICC and from the solid ICC to the y-axis. We hit the value 0.5 on the y-axis, indicating that $\Pr(U_{pi} = 1 \mid \theta_p, \beta_i) = 0.5$.

ICCs are useful for understanding the relationship between ability and probability correct that is predicted by the Rasch model. Following any of the curves in Figure 2.1 from left to right shows an increasing relationship. The smaller values of θ on the left imply smaller values of $\Pr(U_{pi} = 1 \mid \theta_p, \beta_i)$, while the larger values of θ on the right imply larger values of $\Pr(U_{pi} = 1 \mid \theta_p, \beta_i)$. Intuitively, this means that the higher a person's ability, the more likely they are to answer a given item correctly. Since this will be true for every well-designed item of a test, the Rasch model predicts that people with higher ability will tend to answer more test items correctly.

ICCs are also useful for comparing predictions for different items. Let's focus on a single ability $\theta_p = 0$ and compare its probability of a correct response for the three items whose ICCs are shown in Figure 2.1. We see that the probability of correctly responding is highest when $\beta_i = -1$ and lowest when $\beta_i = 1$. This demonstrates the decreasing relationship between difficulty and the predicted probability of correctly responding to an item. Intuitively, this means that the more difficult an item, the less likely a person is to correctly respond to that item. Since this will be true for every person, the Rasch model predicts that more difficult items will be correctly answered by fewer people.

We can get a fuller picture of the effect of difficulty by comparing the locations of the three ICCs in Figure 2.1. We see that the ICC for $\beta_i = -1$ is the farthest to the left, followed by the ICC for $\beta_i = 0$ in the middle and the ICC for $\beta_i = 1$ to the right. This suggests that increasing difficulty shifts the ICC to the right, while decreasing difficulty shifts it to the left. In fact, increasing the difficulty by one effectively increases the ability required to have a particular probability of correctly responding by one.

Now that we understand how difficulty affects an ICC, we can offer a simple interpretation of the difficulty parameter. When $\theta_p = \beta_i$, the probability that test taker p correctly responds to item i is 0.5. Thus, the test taker is as likely to correctly answer the item as not. When $\theta_p > \beta_i$, the test taker is more likely to correctly answer the item. When $\theta_p < \beta_i$, the test taker is less likely to correctly answer the item. This means that we can interpret an item's difficulty as the ability necessary to have a 50-50 chance of correctly answering the item.

From Equation (2.1) and Figure 2.1 we can also see that the probability of test taker p correctly answering item i depends only on the difference between θ_p and β_i, not on their absolute values. The probability of a correct response for a person with an ability of $\theta_p = 2$ for answering an item with a difficulty of $\beta_i = 1$

$$\frac{\exp(2-1)}{1+\exp(2-1)} = \frac{\exp(1)}{1+\exp(1)} \approx 0.73$$

is the same as for a person with an ability of $\theta_p = 1$ and an item with a difficulty $\beta_i = 0$

$$\frac{\exp(1-0)}{1+\exp(1-0)} = \frac{\exp(1)}{1+\exp(1)} \approx 0.73.$$

You can also see this in Figure 2.1 when you read off the probability of a correct response (value on y-axis) for the ICC for $\beta_i = 1$ at the ability $\theta_p = 2$ (value on the x-axis) and compare it to the probability of a correct response for the ICC for $\beta_i = 0$ at the ability $\theta_p = 1$.

2.2.2 Discrimination

Figure 2.2 illustrates an important concept: discrimination. An item's discrimination determines how well the item distinguishes abilities slightly above from abilities slightly below its difficulty. This is typically characterized by the slope of the ICC "in the middle", i.e., for levels of ability close to the item's difficulty, with higher discriminations corresponding to steeper slopes.

As an example, consider the two ICCs in Figure 2.2. Both of the ICCs correspond to items with difficulties of zero. The dashed lines show the probabilities that people with abilities of -0.5 and $+0.5$ correctly solve the item.

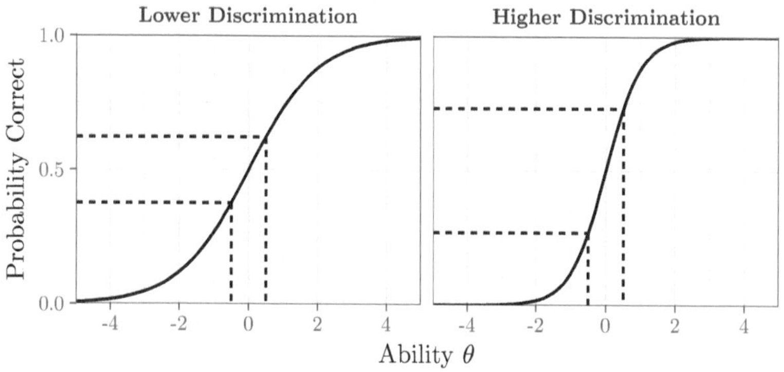

FIGURE 2.2: Demonstration of discrimination. The item for which the ICC is shown on the right has a higher discrimination than the item for which the ICC is shown on the left.

These probabilities are approximately 0.38 and 0.62 for the left item and 0.27 and 0.73 for the right item. This means that two persons with a difference of 1 in their abilities show a difference of 0.24 in their probabilities of solving the left item, and a difference of 0.46 for the right item. The slope in the center of the ICC, where it is virtually linear, can be approximated by this difference on the y-axis divided by the difference on the x-axis as

$$\frac{0.62 - 0.38}{0.5 - (-0.5)} = \frac{0.24}{1} = 0.24$$

for the left item and 0.46 for the right item, respectively.

Thus, the ICC on the right will do a better job than the ICC on the left of discriminating a person whose ability is 0.5 from a person whose ability is -0.5. This means that a person's response to the right item is a better indicator of their ability than their response to the left item, particularly if their ability is close to 0.

An important characteristic of the Rasch model is that it assumes that all items have the same discrimination, specifically a slope of 1. We can see this from its formula, Equation 2.1, where the only parameter for each item i is its difficulty β_i. The formula does not contain an extra parameter for modeling the discrimination. In Section 4.5.1.1, we will present Birnbaum's (1968) two-parameter logistic (2PL) model, which does incorporate a slope parameter. This parameter allows each item to have a different discrimination. As will become clear below, the fact that the Rasch model assumes equal slopes for all items has strong advantages from the measurement point of view. However, it is often considered too rigid from the modeling point of view, because many real tests contain items with unequal discrimination.

2.2.3 The Logistic Function

The Rasch model uses the logistic function to relate ability and difficulty with the probability of a correct response. The logistic function is usually defined[1] as

$$f(x) = \frac{\exp(x)}{1+\exp(x)}.$$

We can obtain the Rasch model by substituting $\theta_p - \beta_i$ for x in this equation. This means that we can compute the probability correct in the Rasch model by applying the logistic function to the difference between θ_p and β_i, i.e., $\Pr(U_{pi} = 1 \mid \theta_p, \beta_i) = f(\theta_p - \beta_i)$.

The logistic function transforms log-odds, or logits, into probabilities. The odds express the relative probability of a correct response versus an incorrect response, and the so-called logit is the natural logarithm of the odds. For example, odds of three indicate that a correct response is three times more likely than an incorrect response, while odds of 1/3 indicate that a correct response is three times less likely than an incorrect response. If the probability of a correct response is given by π, the probability of an incorrect response is $1 - \pi$, and the odds are $O = \frac{\pi}{1-\pi}$.

Any odds O correspond to a unique logit $\log(O)$ and a unique probability π. This allows us to define the Rasch model using the logit, which is an alternative to using the probability, as we have done above. For an ability parameter of θ_p and a difficulty parameter of β_i, it can be shown[2] that the log-odds are $\theta_p - \beta_i$. This leads to the logit form of the Rasch model,

$$\log(O_{pi}) = \log\left\{\frac{\Pr(U_{pi} = 1 \mid \theta_p, \beta_i)}{\Pr(U_{pi} = 0 \mid \theta_p, \beta_i)}\right\} = \theta_p - \beta_i. \tag{2.2}$$

[1]The logistic function is sometimes defined equivalently as

$$f(x) = \frac{1}{1+\exp(-x)},$$

see Section 2.3.3.

[2]We can prove this as follows. As noted above, the odds are defined to be $O = \frac{\pi}{1-\pi}$ and for the Rasch model

$$\pi = \frac{\exp(\theta_p - \beta_i)}{1+\exp(\theta_p - \beta_i)}.$$

As we will derive in detail in Section 2.3.1

$$1 - \pi = 1 - \frac{\exp(\theta_p - \beta_i)}{1+\exp(\theta_p - \beta_i)} = \frac{1}{1+\exp(\theta_p - \beta_i)}.$$

By plugging these terms into the definition of O and canceling the common terms we get

$$O = \frac{\frac{\exp(\theta_p - \beta_i)}{1+\exp(\theta_p - \beta_i)}}{\frac{1}{1+\exp(\theta_p - \beta_i)}} = \exp(\theta_p - \beta_i)$$

and

$$\log(O) = \theta_p - \beta_i.$$

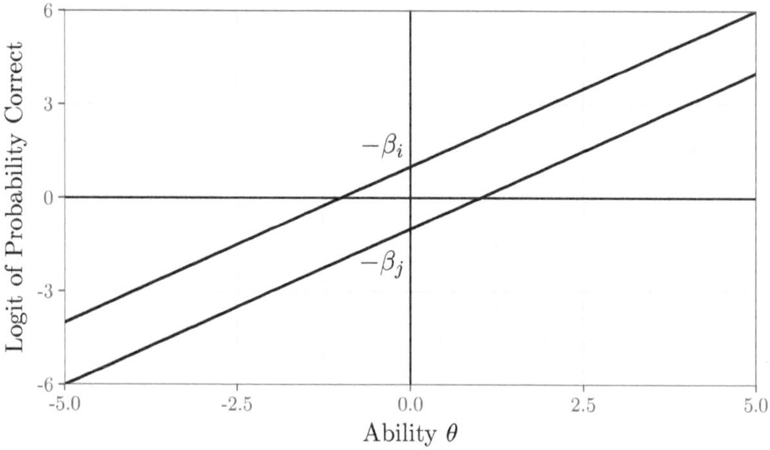

FIGURE 2.3: Effect of difficulty on the logit of the probability of a correct response. The horizontal and vertical lines respectively denote the x and y axes.

Here, $\Pr(U_{pi} = 1 \mid \theta_p, \beta_i)$ corresponds to π and $\Pr(U_{pi} = 0 \mid \theta_p, \beta_i)$ to $1 - \pi$. This way of thinking about the Rasch model will be useful when discussing specific objectivity and when defining the Partial Credit model in Chapter 11.

Another benefit of the logit form is that it offers a simple interpretation of $\theta_p - \beta_i$ as the logit of the probability that person p correctly responds to item i. This means that the logit is linear in θ_p, with y-intercept $-\beta_i$ and slope one. Figure 2.3 demonstrates this for two items i and j with difficulties $\beta_i = -1$ and $\beta_j = 1$. We see that the two lines are parallel, a consequence of having the same slope. The only difference between the lines is where they cross the y-axis: $-\beta_i$, in the case of item i, and $-\beta_j$, in the case of β_j. In this representation, it is directly visible that the ICCs of the Rasch model are parallel, whereas on the probability scale this is only obvious in their middle section (although it can be seen that they do not cross, cf. Figure 2.1).

The fact that the two lines are parallel means that the difference between the logit of the probability of correctly responding to item i and the logit of the probability of correctly responding to item j is the same for every ability. This is one way of understanding specific objectivity, an important property of the Rasch model that we will discuss in Section 2.4.3.

The logistic function is not the only function used in the literature to relate the probability of a correct response to $\theta_p - \beta_i$. Another popular choice is the cumulative distribution function of the standard normal distribution

$$\Phi(x) = \frac{1}{\sqrt{2\pi}} \cdot \int_{-\infty}^{x} \exp(-\tfrac{1}{2}t^2)dt,$$

which links the two quantities in the one-parameter normal ogive (1PNO) model. In the 1PNO, $\Pr(U_{pi} = 1 \mid \theta_p, \beta_i) = \Phi(\theta_p - \beta_i)$. The ICCs of Rasch and 1PNO models look almost the same, but in the 1PNO the term $\theta_p - \beta_i$ does not have an intuitive interpretation in terms of odds like in the Rasch model, and the logistic function is easier to handle mathematically.[3]

2.3 Alternative Representations

It is sometimes convenient to express the Rasch model using equations other than Equation (2.1). We present these different representations here so that you recognize them when they are used in later chapters or in other articles or books. We start by working out an expression for the probability of an incorrect response, $\Pr(U_{pi} = 0 \mid \theta_p, \beta_i)$.

2.3.1 Probability of an Incorrect Response

The Rasch model does not allow for partial credit. Thus, U_{pi} is either zero or one for every person and test item, and the probabilities of these two possible outcomes must sum up to one, which means

$$\Pr(U_{pi} = 1 \mid \theta_p, \beta_i) + \Pr(U_{pi} = 0 \mid \theta_p, \beta_i) = 1.$$

Solving for $\Pr(U_{pi} = 0 \mid \theta_p, \beta_i)$, we get

$$\Pr(U_{pi} = 0 \mid \theta_p, \beta_i) = 1 - \Pr(U_{pi} = 1 \mid \theta_p, \beta_i) = 1 - \frac{\exp(\theta_p - \beta_i)}{1 + \exp(\theta_p - \beta_i)},$$

after substituting Equation (2.1) for $\Pr(U_{pi} = 1 \mid \theta_p, \beta_i)$. We can simplify the last expression by expanding the 1 to obtain

$$\begin{aligned}\Pr(U_{pi} = 0 \mid \theta_p, \beta_i) &= \frac{1 + \exp(\theta_p - \beta_i)}{1 + \exp(\theta_p - \beta_i)} - \frac{\exp(\theta_p - \beta_i)}{1 + \exp(\theta_p - \beta_i)} \\ &= \frac{1 + \exp(\theta_p - \beta_i) - \exp(\theta_p - \beta_i)}{1 + \exp(\theta_p - \beta_i)} \\ &= \frac{1}{1 + \exp(\theta_p - \beta_i)}.\end{aligned}$$

[3]You will sometimes see a scaling factor of 1.7 being used to compute back and forth between the logistic and normal ogive model, because this factor minimizes the distance between the two functions.

2.3.2 Probability of an Arbitrary Response

In this section, we will work out the probability of an arbitrary response $\Pr(U_{pi} = u_{pi} \mid \theta_p, \beta_i)$, which combines separate expressions for $\Pr(U_{pi} = 1 \mid \theta_p, \beta_i)$ and $\Pr(U_{pi} = 0 \mid \theta_p, \beta_i)$ into one. Having a single expression is useful for writing down probabilities of response patterns and entire data matrices. For the Rasch model, this expression is

$$\Pr(U_{pi} = u_{pi} \mid \theta_p, \beta_i) = \frac{\exp\{u_{pi} \cdot (\theta_p - \beta_i)\}}{1 + \exp(\theta_p - \beta_i)}, \tag{2.3}$$

where it is still open whether u_{pi} will take the value zero or one.

We can verify that the two separate equations for both cases are covered by Equation (2.3) by means of substituting each of the two possible values of u_{pi}. When $u_{pi} = 1$, the expression reduces to

$$\frac{\exp\{1 \cdot (\theta_p - \beta_i)\}}{1 + \exp(\theta_p - \beta_i)} = \frac{\exp(\theta_p - \beta_i)}{1 + \exp(\theta_p - \beta_i)}.$$

When $u_{pi} = 0$ it gives

$$\frac{\exp\{0 \cdot (\theta_p - \beta_i)\}}{1 + \exp(\theta_p - \beta_i)} = \frac{\exp(0)}{1 + \exp(\theta_p - \beta_i)} = \frac{1}{1 + \exp(\theta_p - \beta_i)},$$

since $\exp(0) = 1$, verifying Equation (2.3).

In some literature, particularly in the literature on Bayesian IRT, a different combined expression is used. This expression is based on the Bernoulli distribution, named after the Swiss mathematician Jacob Bernoulli. The Bernoulli distribution gives the probabilities of any random variable with two outcomes, like responses in the Rasch model. It has a single parameter, giving the probability that the random variable is one. For the Rasch model, this parameter is

$$\pi_{pi} = \Pr(U_{pi} = 1 \mid \theta_p, \beta_i) = \frac{\exp(\theta_p - \beta_i)}{1 + \exp(\theta_p - \beta_i)}.$$

Given π_{pi}, the probability of response u_{pi} is

$$\Pr(U_{pi} = u_{pi} \mid \theta_p, \beta_i) = \pi_{pi}^{u_{pi}} \cdot (1 - \pi_{pi})^{1 - u_{pi}}. \tag{2.4}$$

The base of the second term, $1 - \pi_{pi}$, is equal to the probability of an incorrect response $\Pr(U_{pi} = 0 \mid \theta_p, \beta_i)$. Thus, the probability of a correct response is raised to the power u_{pi}, while the probability of an incorrect response is raised to the power $1 - u_{pi}$.

This expression is really just a notational trick exploiting the fact that $a^0 = 1$ for any a. The u_{pi} and $1 - u_{pi}$ act as switches. When $u_{pi} = 1$, $1 - u_{pi} = 0$, so the probability of a correct response is turned on while the probability of an incorrect response is switched off by being set to one. We can watch this play out in the equations by substituting 1 for u_{pi}. This gives

$$\pi_{pi}^1 \cdot (1 - \pi_{pi})^{1-1} = \pi_{pi}^1 \cdot (1 - \pi_{pi})^0 = \pi_{pi} \cdot 1 = \pi_{pi}.$$

Setting $u_{pi} = 0$ reverses the switches: the probability of a correct response is turned off, while the probability of an incorrect response is turned on.

A benefit of using Equation (2.4) for describing the Rasch model is that it does not depend on the specific form of the ICC. For example, we could obtain a unified equation for the response probabilities in the 1PNO model by defining π_{pi} to be the probability given by the 1PNO instead of the one given by the Rasch model. A second benefit is that it allows us to use statistical shorthand to hide its mathematical details in many situations. When we want to say that the random variable U_{pi} follows a Bernoulli distribution with parameter π_{pi}, it is common to write $U_{pi} \sim \text{Bernoulli}(\pi_{pi})$. The "$\sim$" can be read as "is distributed as". In this notation, which is common in Bayesian IRT, we can write the Rasch model as

$$\begin{aligned} U_{pi} &\sim \text{Bernoulli}(\pi_{pi}), \\ \pi_{pi} &= \frac{\exp(\theta_p - \beta_i)}{1 + \exp(\theta_p - \beta_i)}. \end{aligned}$$

2.3.3 Alternative Representation of the Logistic Function

The logistic function used in the Rasch model can be written in two ways: With the exponential function in both numerator and denominator, like we do in most parts of the book (left), or equivalently with the exponential function only in the denominator, followed by its negative argument (right):

$$\frac{\exp(\theta_p - \beta_i)}{1 + \exp(\theta_p - \beta_i)} = \frac{1}{1 + \exp\left(-(\theta_p - \beta_i)\right)}$$

We only list this formulation here to avoid confusion in case you come across it in other books.

2.3.4 Multiplicative Form

Finally, some authors also make use of a multiplicative form of the Rasch model. The multiplicative Rasch model uses a property of the exponential function to split each $\exp(\theta_p - \beta_i)$ into $\exp(\theta_p) \cdot \exp(-\beta_i)$. It then replaces $\exp(\theta_p)$ with $\xi_p = \exp(\theta_p)$ and $\exp(-\beta_i)$ with $\sigma_i = \exp(-\beta_i)$, resulting in

$$\Pr(U_{pi} = 1 \mid \xi_p, \sigma_i) = \frac{\xi_p \cdot \sigma_i}{1 + \xi_p \cdot \sigma_i}.$$

Intuitively, we might think of the multiplicative form as an "odds" form of the Rasch model. From Section 2.2.3, we know that $\exp(\theta_p - \beta_i)$ corresponds to the odds of a correct response. The multiplicative form of the Rasch model expresses the odds of a correct response using the product of a person component ξ_p and an item component σ_i.

2.4 Properties and Assumptions

In this section, we review important properties of the Rasch model. These properties are the reason why the Rasch model is so theoretically appealing and have led to its widespread use. Perhaps the most appealing of these properties is the fact that it allows for objective measurement of latent traits.

Didactically, it is easier to start with the Rasch item response function and show that the various properties of the Rasch model follow. However, it turns out the Rasch model can be mathematically derived from collections of these properties, typically called assumptions or axioms (cf., e.g., Fischer & Molenaar, 1995). These assumptions need to be checked whenever we apply the Rasch model to real data.

We encountered an example of this in Section 2.2.2. There we noted that the Rasch model assumes that every item has equal discrimination, resulting in parallel ICCs. When applying the Rasch model to real data, we will need to check this assumption (see Chapter 4). If it does not hold empirically, then we cannot apply the Rasch model to this set of test items. As we have already outlined, we then either need to exclude those items that are not in line with the Rasch measurement model, or use a more flexible model to better describe the given items.

In the following, as well as in Chapter 4, we will also note that the assumptions are not fully separable, both theoretically and with respect to empirical violations of the Rasch model.

2.4.1 Sufficient Statistics

A statistic is defined to be any function of observed data. Typically, statistics are used to summarize important features of data. The sample mean over the values of x_p for the persons $1, \ldots, P$

$$\bar{x} = \frac{1}{P} \sum_{p=1}^{P} x_p$$

for example, summarizes the information contained in the sample about the mean of a population, the expected value. That is why it is often used to

	Item						
Person	1	2	3	4	5	6	r_p
1	0	1	0	1	0	1	3
2	0	1	1	0	1	1	4
3	0	1	1	1	0	0	3
4	1	0	0	1	0	0	2
c_i	1	3	2	3	1	2	

TABLE 2.3: Data matrix of a test with row sums and column sums.

estimate the expected value from samples. Statistically, we say that $\bar{x}$ is an estimator for the expected value.

$\bar{x}$ is not the only statistic that we could use to estimate a population mean. For example, we might try the mean of the first, third, and fifth value, $x^* = \frac{1}{3}(x_1 + x_3 + x_5)$, instead of $\bar{x}$. This statistic would also estimate the population mean, but it would be less accurate than $\bar{x}$, on average. The reason is that x^* ignores the information about the population mean carried by x_2, x_4, etc.

It turns out that $\bar{x}$ extracts *all* of the information about the population mean contained in the sample. Statistics with this property are called sufficient statistics . This contrasts with statistics like x^*, that ignore some of the information about the population mean. Since sufficient statistics contain all of the information in the sample about a quantity of interest, knowing the sufficient statistic makes the individual samples irrelevant. That is the beauty of sufficient statistics: they extract all of the information in a sample about a quantity of interest and summarize it using a single number.

Every unknown parameter in the Rasch model has a sufficient statistic. The number of items answered correctly by person p, or sum score, is a sufficient statistic for the person parameter θ_p. We can compute person p's sum score by taking the sum of all of the entries in row p of the data matrix. For this reason, the sum scores are often called row sums, and we will denote them using r_p. Similarly, the number of people correctly answering item i is a sufficient statistic for the item parameter β_i. We can compute this statistic by taking the sum of all of the entries in column i. We will denote the sum of column i using c_i. The row and column sums for our introductory example are shown in Table 2.3.

The fact that a person's sum score is a sufficient statistic for the person parameter θ_p means that we do not need to know which items a person has solved to estimate their ability, only how many items they solved. At first glance, this appears to contradict our requirement that the probability of a correct response should depend on both the ability of the person and the difficulty of the item. If a person's responses to test items depend on both their ability and the difficulties of the test items, how can the row sums contain all of the information about a person's ability? Shouldn't the difficulties of the individual items matter? However, if we look at the pattern of correct and incorrect answers of a person across an entire test, it makes sense that the total number of items a person solves would be indicative of his or her ability. A person with low ability will only be able to solve easy items, while a person with high ability will be able to solve easy items as well as harder items. Consequently, we would expect people with high ability to answer more items correctly in total.

This does not mean that people with higher ability will always correctly answer more items correctly. The Rasch model describes the relationship between a person's ability and their test responses probabilistically. This probabilistic approach was an important advancement of the Rasch model and related models over the deterministic view underlying the perfect response patterns of a Guttman structure (described in detail, e.g., in Andrich & Marais, 2019). For this reason, we cannot say for certain how a person will respond to an individual item, even when we know the ability of the person and the difficulty of the item. As discussed in Section 2.2, many factors besides ability can influence a person's response on a particular test administration. The Rasch model does not take these into account explicitly, but subsumes them in the probabilistic response. In Chapter 10, models that explicitly account for additional aspects, such as guessing in multiple choice items, are introduced.

The Rasch model states that on average a person with a higher ability is more likely to correctly answer an item of a given difficulty. To illustrate this point, imagine a person with higher ability and a person with lower ability. Each completes 100 different items of the same medium difficulty. The person with higher ability will likely answer not all but, for example, about 80 of those 100 items correctly, and the person with lower ability not none but about 20, because the answering is not deterministically but probabilistically driven by the ability and the difficulty.

2.4.2 Local Stochastic Independence

Most statistical models assume that individual events are stochastically independent of one another. This means that knowing how one of the events turns out gives us no information about how any of the other events turns out. Another way of saying this is that knowing how one of the events turns out does not change the probabilities of any of the other events. In practical applications, this assumption is usually reasonable, and it makes computing the joint probability of all of the events much easier. For example, suppose we would like to calculate the probability that flipping a coin twice yields two heads. For a fair coin, there is a 50-50 chance of flipping heads on each flip. Knowing that the first flip landed on heads does not change this probability.

The joint probability of two independent events is equal to the product of their individual probabilities. We can use this fact to compute the probability of two heads in the coin flip example. Since each flip has a 0.5 probability of landing on heads, the probability of both flips landing on heads is $0.5 \times 0.5 = 0.25$.

2.4.2.1 Items

The Rasch model uses this assumption to define the joint probability of a person's test responses given their ability and the difficulties of the items. As in the coin flipping example, we can compute the probability that person p correctly responds to the first two items of a test by multiplying $\Pr(U_{p1} = 1 \mid \theta_p, \beta_1)$ by $\Pr(U_{p2} = 1 \mid \theta_p, \beta_2)$. We can obtain the joint probability of an arbitrary pattern of responses in two items with item difficulties β_1 and β_2 in the same way, meaning

$$\begin{aligned}\Pr(U_{p1} = u_{p1}, U_{p2} = u_{p2} \mid \theta_p, \beta_1, \beta_2) = \Pr(U_{p1} = u_{p1} \mid \theta_p, \beta_1)\\ \times \Pr(U_{p2} = u_{p2} \mid \theta_p, \beta_2).\end{aligned}$$

Clearly, this would become cumbersome to write out for a test of I items. We can simplify things using vectors. A vector is simply an indexed collection of numbers. For example, the vector $\boldsymbol{\beta} = (\beta_1, ..., \beta_I)$ collects all of the item parameters, and the vector $\boldsymbol{u}_{p\cdot} = (u_{p1}, ..., u_{pI})$ collects all of person p's responses to the test items. The ith elements of these vectors are the item parameter β_i of the ith item and the response u_{pi} of person p to item i, respectively.

Random vectors extend this idea to random variables. A random vector is an indexed collection of random variables. For example, we can collect all of the random variables associated with person p's responses to the I items into a random vector $\boldsymbol{U}_{p\cdot} = (U_{p1}, ..., U_{pI})$. The ith element of $\boldsymbol{U}_{p\cdot}$ is U_{pi}, the random variable associated with person p's response to item i. This allows us to simplify

$$\Pr(U_{p1} = u_{p1}, ..., U_{pI} = u_{pI} \mid \theta_p, \beta_1, ..., \beta_I) = \Pr(\boldsymbol{U}_{p\cdot} = \boldsymbol{u}_{p\cdot} \mid \theta_p, \boldsymbol{\beta})$$

using the fact that two vectors are equal whenever all of their elements are equal.

We can use random vectors to work out the joint probability of person p's pattern of responses to the I test items using the product symbol $\prod_{i=1}^{I}$. The probability that person p gives response pattern $\boldsymbol{u}_{p\cdot}$ is

$$\Pr(\boldsymbol{U}_{p\cdot} = \boldsymbol{u}_{p\cdot} \mid \theta_p, \boldsymbol{\beta}) = \prod_{i=1}^{I} \Pr(U_{pi} = u_{pi} \mid \theta_p, \beta_i),$$

when all of p's responses are independent. Substituting the righthand side of Equation (2.3) for $\Pr(U_{pi} = u_{pi} \mid \theta_p, \beta_i)$, we get

$$\Pr(\boldsymbol{U}_{p\cdot} = \boldsymbol{u}_{p\cdot} \mid \theta_p, \boldsymbol{\beta}) = \prod_{i=1}^{I} \frac{\exp\{u_{pi}(\theta_p - \beta_i)\}}{1 + \exp(\theta_p - \beta_i)} = \frac{\prod_{i=1}^{I} \exp\{u_{pi}(\theta_p - \beta_i)\}}{\prod_{i=1}^{I}[1 + \exp(\theta_p - \beta_i)]}$$

after switching the order of multiplication and division. Focusing only on the numerator for now, we push the product into the exponential[4] to obtain

$$\prod_{i=1}^{I} \exp\{u_{pi} \cdot (\theta_p - \beta_i)\} = \exp\left\{\sum_{i=1}^{I} [u_{pi} \cdot (\theta_p - \beta_i)]\right\}.$$

We can further simplify the numerator by distributing u_{pi} to obtain

$$\begin{aligned}\exp\left\{\sum_{i=1}^{I} [u_{pi} \cdot (\theta_p - \beta_i)]\right\} &= \exp\left\{\sum_{i=1}^{I} [u_{pi} \cdot \theta_p - u_{pi} \cdot \beta_i]\right\} \\ &= \exp\left\{\sum_{i=1}^{I} u_{pi} \cdot \theta_p - \sum_{i=1}^{I} u_{pi} \cdot \beta_i\right\}.\end{aligned}$$

The first sum can be simplified as

$$\sum_{i=1}^{I} u_{pi} \cdot \theta_p = \theta_p \cdot \sum_{i=1}^{I} u_{pi},$$

since θ_p does not depend on i. Moreover, $\sum_{i=1}^{I} u_{pi}$ is the number of items that were correctly answered by person p, which we have denoted r_p. Thus,

$$\theta_p \cdot \sum_{i=1}^{I} u_{pi} = \theta_p \cdot r_p.$$

Substituting these results back into the full expression for $\Pr(\boldsymbol{U}_{p\cdot} = \boldsymbol{u}_{p\cdot} \mid \theta_p, \boldsymbol{\beta})$, we obtain

$$\Pr(\boldsymbol{U}_{p\cdot} = \boldsymbol{u}_{p\cdot} \mid \theta_p, \boldsymbol{\beta}) = \frac{\exp\{\theta_p \cdot r_p - \sum_{i=1}^{I} u_{pi} \cdot \beta_i\}}{\prod_{i=1}^{I}[1 + \exp(\theta_p - \beta_i)]}. \tag{2.5}$$

We will come back to this equation in the chapter on parameter estimation.

It only makes sense to multiply the probabilities of events when it is reasonable to assume the events are independent. For the Rasch model, this means that the probabilities of solving the different test items only depend on one another through the ability of the person. For a given ability, knowing that the first item was answered correctly would not give us any information about whether the second item was answered correctly, etc. The term "local" in "local stochastic independence" refers to the fact that the independence of items is required locally, i.e., for a given ability, and that the independence of persons is required for a given item difficulty. Sometimes, the term local stochastic independence is also used to refer to only one of these two requirements at a time, for instance when we describe tests for this assumption in 4.3.5.

---

[4]This is a general property of the exponential function: $\exp(x) \cdot \exp(y) = \exp(x+y)$ (see Appendix A).

One situation where this assumption is unreasonable is when the response to a test item depends on a person's response to an earlier item. This happens, e.g., in mathematics tests where the calculations required to solve one item depend on calculations previously done to solve another. For example, the probability that a person correctly solves the second item depends on whether the person correctly solved the first item, since the solution to the first item is required to solve the second. This demonstrates why it is important to formulate test items so that they can be solved without the answers to other items. Without local stochastic independence, the Rasch model cannot be applied.

The local independence assumption is also violated for testlets. A testlet is a group of items sharing a common theme. For example, many standardized tests contain groups of typically two to five items asking about the same passage of text. In this case, responses to the items about the passage will be related through the person's understanding of the passage. Knowing the person's responses to one or more items offers information about how well he or she understood the passage, which affects their probability to answer the remaining items on the same passage correctly, even given their ability. Thankfully, testlets are dependent by design and their dependence structure can be accounted for using testlet models (e.g., Wainer, Bradlow, & Wang, 2007).[5]

2.4.2.2 Persons

We can extend the joint probability for a single person's responses to the joint probability of all of the test takers' responses by assuming that all of the test takers' responses are independent. For notational convenience, we collect the test takers' ability parameters θ_p into a vector $\boldsymbol{\theta}$ whose pth element is the ability parameter of person p. We also stack the $\boldsymbol{U}_{p\cdot}$ to form a random matrix $\boldsymbol{U}$. Each row of the matrix contains the random variables associated with the responses of a single person. Each column contains the random variables associated with all responses to a single test item. Equally, we stack the actual responses from each person to form a matrix $\boldsymbol{u}$. The entry in row p, column i is u_{pi}, i.e., the response of person p to item i.

This allows us to derive the joint probability of all of the test takers' responses to all of the test items. This is

$$\begin{aligned}\Pr(\boldsymbol{U} = \boldsymbol{u} \mid \boldsymbol{\theta}, \boldsymbol{\beta}) &= \Pr(\boldsymbol{U}_{1\cdot} = \boldsymbol{u}_{1\cdot}, \ldots, \boldsymbol{U}_{P\cdot} = \boldsymbol{u}_{P\cdot} \mid \theta_1, \ldots, \theta_P, \boldsymbol{\beta}) \\ &= \prod_{p=1}^{P} \frac{\exp\{r_p \cdot \theta_p - \sum_{i=1}^{I} u_{pi} \cdot \beta_i\}}{\prod_{i=1}^{I}[1 + \exp(\theta_p - \beta_i)]},\end{aligned}$$

[5]Note that the term testlet is also used to describe an approach where the responses to locally dependent items are aggregated to form one polytomous "super item" (e.g., Keller, Swaminathan, & Sireci, 2003). These super items are then analyzed by means of polytomous models (cf. Chapter 11).

after substituting our result for the probability of a single person's responses. We again push the product into the numerator and denominator and pull the product in the numerator into the exponential, yielding

$$\begin{aligned}\Pr(\boldsymbol{U} = \boldsymbol{u} \mid \boldsymbol{\theta}, \boldsymbol{\beta}) &= \frac{\prod_{p=1}^{P} \exp\{r_p \cdot \theta_p - \sum_{i=1}^{I} u_{pi} \cdot \beta_i\}}{\prod_{p=1}^{P} \prod_{i=1}^{I} [1 + \exp(\theta_p - \beta_i)]} \\ &= \frac{\exp\{\sum_{p=1}^{P} (r_p \cdot \theta_p - \sum_{i=1}^{I} u_{pi} \cdot \beta_i)\}}{\prod_{p=1}^{P} \prod_{i=1}^{I} [1 + \exp(\theta_p - \beta_i)]}.\end{aligned}$$

Focusing only on the numerator again, we can distribute the sum across the θ_p and β_i terms to obtain

$$\Pr(\boldsymbol{U} = \boldsymbol{u} \mid \boldsymbol{\theta}, \boldsymbol{\beta}) = \frac{\exp\{\sum_{p=1}^{P} r_p \cdot \theta_p - \sum_{p=1}^{P} \sum_{i=1}^{I} u_{pi} \cdot \beta_i\}}{\prod_{p=1}^{P} \prod_{i=1}^{I} [1 + \exp(\theta_p - \beta_i)]}.$$

We can switch the summation order in the double sum, since

$$\sum_{p=1}^{P} \sum_{i=1}^{I} u_{pi} \cdot \beta_i = \sum_{i=1}^{I} \sum_{p=1}^{P} u_{pi} \cdot \beta_i.$$

The β_i terms do not depend on p, so we can pull them out of the inner sum. Moreover, $\sum_{p=1}^{P} u_{pi}$ is the sum of column i, which we have denoted c_i. Thus,

$$\sum_{p=1}^{P} \sum_{i=1}^{I} u_{pi} \cdot \beta_i = \sum_{i=1}^{I} \sum_{p=1}^{P} u_{pi} \cdot \beta_i = \sum_{i=1}^{I} c_i \cdot \beta_i.$$

Pulling this back into the full expression for $\Pr(\boldsymbol{U} = \boldsymbol{u} \mid \boldsymbol{\theta}, \boldsymbol{\beta})$ gives

$$\Pr(\boldsymbol{U} = \boldsymbol{u} \mid \boldsymbol{\theta}, \boldsymbol{\beta}) = \frac{\exp\{\sum_{p=1}^{P} r_p \cdot \theta_p - \sum_{i=1}^{I} c_i \cdot \beta_i\}}{\prod_{p=1}^{P} \prod_{i=1}^{I} [1 + \exp(\theta_p - \beta_i)]},$$

the desired probability.[6]

Just as we must check whether the assumption that a single person's responses to different items are independent is reasonable, we must also check whether the responses of different people are independent. This means that knowing whether one person correctly answered a given item should not give us any information about whether a second person correctly answered the item. One situation where this is clearly violated is when one person cheats by copying answers from another.

[6] As an interesting sidenote, the existence of this expression proves the sufficiency of the row and column sums through the properties of so-called exponential families. We refer the interested reader to Casella and Berger (2002).

The reasonableness of the local independence assumption should always be checked before applying the Rasch model (see Chapter 4). When this assumption is reasonable, it simplifies calculating the joint probability of the test responses. Moreover, when combined with the sufficiency of the row sums and a few other minor assumptions, local stochastic independence *implies* the Rasch model item response function. An early derivation of the Rasch model based on these assumptions is known as Andersen's theorem (e.g., Fischer & Molenaar, 1995; McDonald, 2011).

2.4.3 Specific Objectivity

The fundamental purpose of psychometric tests is to compare individuals, and ensuring fairness is paramount in these comparisons. Specific objectivity ensures an aspect of fairness by requiring that comparisons between people only depend on their respective abilities and not on the specific items used to compare them. It states that, if one person has a higher probability than another of answering correctly for one item, then they must have a higher probability of answering correctly for every item. Specific objectivity also applies to items. In this case, it states that if one item is easier than another for one person, then it must be easier than another for every person. An intuitive way to check whether specific objectivity is fulfilled by a model is to ensure that the theoretical ICCs of different items do not cross. This is the case for the Rasch model, cf. Figure 2.1. We will show how crossing ICCs would violate specific objectivity below.

Some sources (e.g., Irtel, 1996) define specific objectivity more strictly algebraically and require that the ratio between the odds of correct responses, or odds ratio, for any two people p and q must be the same for every test item. Let O_{pi} be the odds that person p correctly responds to item i. Then, specific objectivity requires

$$\frac{O_{p1}}{O_{q1}} = \ldots = \frac{O_{pI}}{O_{qI}}.$$

This means that for any pair of items i and j, the odds ratio between two persons p and q for item i is equal to that of item j, i.e., $O_{pi}/O_{qi} = O_{pj}/O_{qj}$, which is equivalent to the condition that $O_{pi} \cdot O_{qj} = O_{pj} \cdot O_{qi}$. Thus, the requirement that the odds ratio be constant across items is also known as the multiplication condition.

Alternatively, specific objectivity can be stated in terms of test items. In this case, it requires that the odds ratio between any two items be the same for every person, i.e.,

$$\frac{O_{1i}}{O_{1j}} = \ldots = \frac{O_{Pi}}{O_{Pj}}.$$

These expressions are equivalent. If the multiplication condition holds for people, then it holds for test items, and vice versa.[7]

We can better understand the multiplication condition by example. Suppose that Marco and Cora complete a twenty-item test. Marco has a probability of $0.2 = 1/5$ for solving the first item. Therefore, the probability that Marco does not solve it is $1 - 0.2 = 0.8 = 4/5$. Overall, his odds for solving it are $0.2 : 0.8 = 1 : 4$. On the other hand, Cora has a probability of 0.5 of solving this item, which corresponds to $1 : 1$ odds. We denote Marco's odds for the first item by $O_{\text{Marco},1}$ and Cora's by $O_{\text{Cora},1}$. Using this notation, we now obtain

$$\frac{O_{\text{Cora},1}}{O_{\text{Marco},1}} = \frac{1/1}{1/4} = 4,$$

meaning that the odds that Cora correctly answers the first item are four times as large as the odds that Marco correctly answers the first item. Now let $O_{\text{Cora},i}$ and $O_{\text{Marco},i}$ denote the corresponding odds for any arbitrary item i. Specific objectivity requires that

$$\frac{O_{\text{Cora},i}}{O_{\text{Marco},i}} = 4$$

for each of the nineteen remaining items. This means that, if the odds that Marco correctly answers the third item are, e.g., $1 : 1$, the odds that Cora correctly answers the third item have to be $4 : 1$. This provides us with a specific criterion for checking specific objectivity in this strict sense: It requires that this relationship holds for every pair of test takers.

Suppose that a third person, Jo, completes the test. Her odds of correctly answering the first item are $3 : 2$. If we compare her odds with those of Marco and Cora, she is six times more likely to correctly answer the first item than Marco and 1.5 times more likely than Cora. Then, specific objectivity in the strict sense requires that this ratio must also hold for the other items. Therefore, the odds that Cora correctly answers any test item i must be four times the corresponding odds for Marco. Further, the odds that Jo correctly answers any test item i must be six times the corresponding odds for Marco and 1.5 times the corresponding odds for Cora. Using our formal notation, we get

$$\begin{aligned} O_{\text{Cora},i} &= 4 \times O_{\text{Marco},i} \\ O_{\text{Jo},i} &= 6 \times O_{\text{Marco},i} \\ O_{\text{Jo},i} &= 1.5 \times O_{\text{Cora},i} \end{aligned}$$

for every test item i.

[7] The person-centric version of the multiplication condition states that

$$\frac{O_{pi}}{O_{qi}} = \frac{O_{pj}}{O_{qj}}$$

for all pairs of people (p, q) and all pairs of items (i, j). If we multiply both sides by O_{qi} and divide both sides by O_{pj}, then we get the item-centric version of the multiplication condition.

At an intuitive level, we demonstrated that the Rasch model satisfies specific objectivity in Section 2.2.3 by illustrating that the graphs of the relationship between ability and the logit of the probability of a correct response form parallel lines. We can show this again formally by taking the logs of the person and item multiplication conditions to obtain difference conditions on the logits. In particular, we use the fact that $\log(x/y) = \log(x) - \log(y)$ for two real numbers x and y (see Appendix A):

$$\log(O_{p1}) - \log(O_{q1}) = \ldots = \log(O_{pI}) - \log(O_{qI})$$

and

$$\log(O_{1i}) - \log(O_{1j}) = \ldots = \log(O_{Pi}) - \log(O_{Pj}),$$

respectively. We know from Section 2.2.3 that for the Rasch model the logit $\log(O_{pi}) = \theta_i - \beta_j$. Thus,

$$\log(O_{pi}) - \log(O_{qi}) = (\theta_p - \beta_i) - (\theta_q - \beta_i) = \theta_p - \theta_q,$$

since the β_is cancel out. As $\theta_p - \theta_q$ does not depend on i, the person multiplication condition holds. Moreover,

$$\log(O_{pi}) - \log(O_{pj}) = (\theta_p - \beta_i) - (\theta_p - \beta_j) = \beta_j - \beta_i,$$

since the θ_ps cancel out. As $\beta_j - \beta_i$ does not depend on p, the item multiplication condition holds. Thus, the Rasch model satisfies specific objectivity.

We demonstrate why any model allowing for different item discriminations violates specific objectivity using Figure 2.4. Consider the items whose ICCs are shown in the right of the figure. As opposed to those items in the left panel, the ones in the right panel have different discriminations, as demonstrated by the fact that their ICCs cross. Looking closer, we see that person p is more likely to correctly answer the item with the solid ICC than the item with the dashed ICC, while person q is more likely to correctly answer the item with the dashed ICC than the item with the solid ICC. Comparing the two items using person p, we get that

$$\log(O_{p,\text{solid}}) - \log(O_{p,\text{dashed}}) > 0,$$

as p is more likely to answer the solid item. Comparing the two items using person q, we get that

$$\log(O_{q,\text{solid}}) - \log(O_{q,\text{dashed}}) < 0,$$

since q is more likely to answer the dashed item correctly. Since these two differences are not equal, the multiplication condition does not hold. More generally speaking, since for one person one item is easier, but for the other person the other item is easier, specific objectivity is violated whenever ICCs are not parallel or even cross.

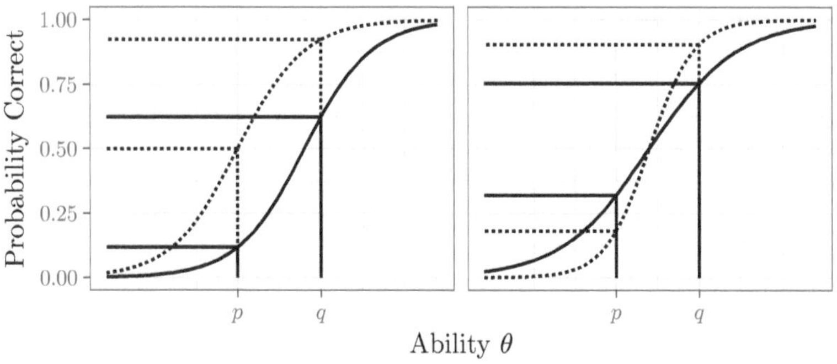

FIGURE 2.4: Demonstration that crossing ICCs violate specific objectivity.

In Figure 2.4, the comparison of items depended on the ability of the people. However, the comparison of items can also depend on other factors. For example, an item testing mathematical ability containing a particularly complicated or uncommon phrase in its task description will, compared to the other items, be more difficult for students who are not native English speakers, regardless of their mathematical ability. Items that are differentially difficult for different groups of test takers are said to exhibit differential item functioning (DIF).[8] An item with DIF is not suitable for comparing different groups of people, so it needs to be modified or removed from the test. Statistical tests for detecting DIF will be covered in Chapter 4.

Specific objectivity is sometimes also referred to as "sample independence". Unfortunately, this terminology can be misleading. Suppose, for example, that we have developed a questionnaire for measuring leadership capacity and have tested its Rasch scalability on a group of investment bankers. Thinking of specific objectivity as sample independence might lead to the mistaken conclusion that the questionnaire can be transferred directly to software engineers or flight attendants. This type of conclusion cannot be justified using specific objectivity. Investment bankers might interpret the questionnaire items differently than software engineers and flight attendants. This could lead to DIF and violate specific objectivity. We need to determine empirically whether the Rasch model, and specific objectivity, make sense for each new group of people. Rasch himself argued this point in a 1965 lecture cited by Gustafsson (1980, p. 231): "In an empirical science specific objectivity can never be fully ascertained if the objects and/or agents is an infinite set; it can only be set up as a working hypothesis which has got to be carefully tested [...]. And whenever additional data are collected we must be ready to do it over again [...]."

[8]If the difference in difficulty between the groups is the same over all abilities, the DIF is called uniform. If it varies with the ability, it is called nonuniform.

2.4.4 Unidimensionality

The Rasch model assumes that people can be ordered on a single latent dimension. This is exemplified by the fact that each person is assigned exactly one ability. Conceptually, saying that a mathematics test is "unidimensional" means that it only measures mathematical ability. A test like the SAT would not be unidimensional because it tests both mathematical and verbal ability. For example, one person might score above another on mathematical ability, but below on verbal ability. Individually, however, the mathematics and verbal sections are considered unidimensional.

Unaccounted multidimensionality can also be the source of DIF. For example, a person's score on a math test can also depend on their language ability to some degree, since it influences their ability to read and respond to the test items. Note that in this example as opposed to the previous one, language ability is not a primary dimension of interest, but a secondary or nuisance dimension. If two groups of test takers differ in their distribution for the secondary dimension, this can induce DIF in items measuring this secondary dimension in addition to the primary dimension (for details see Ackerman, 1992; Roussos & Stout, 1996). As a result, statistical tests assessing the presence of DIF can also be sensitive to multidimensionality. Such tests will be discussed in Chapter 4.

Additionally, many psychological constructs are inherently multidimensional. For example, Carroll (1993) argues for a hierarchically structured model dividing intelligence in many different ways (e.g., fluid and crystallized intelligence). While the unidimensional Rasch model cannot account for multidimensional traits and abilities, we will point to multidimensional extensions in Section 10.3 that can.

2.4.5 Measurement Scale

Throughout this chapter, we have described the Rasch model as a tool for measuring ability. What does it mean to "measure" ability? In a seminal paper, Stevens (1946) defines measurement as "the assignment of numerals to objects or events according to rules". Stevens' sweeping definition enjoys nearly universal acceptance in psychology (though arguments against it can be found, e.g., Michell, 1999). It is broad enough to encompass familiar modes of physical measurement, such as using a ruler to measure distance, as well as psychological measurement, such as experimental paradigms for measuring perceived loudness.

Stevens divides measurement paradigms into four classes, known as measurement scales. The first of these, nominal scales, use numbers merely as labels, in the same way football players are numbered to distinguish them from one another. The second, ordinal scales, use numbers to order. The ranks assigned to athletes in the Olympics, for example, are an ordinal scale. The final two scales, interval and ratio, allow for relative distances. The difference between these scales is that ratio scales have a natural zero point, while interval scales do not. To illustrate this, let's consider scales for temperature and length as examples of interval and ratio scales, respectively. Different units for temperature have different zero points. For example, 0 degrees Celsius is about 32 degrees Fahrenheit. Length, by contrast, has a natural zero point: 0 meters is the same as 0 feet or 0 miles, namely no length at all.

Before moving to the Rasch model, let's consider the temperature units Celsius and Fahrenheit in a bit more detail. Any Fahrenheit temperature F can be converted into a Celsius temperature C by first subtracting 32 from F and then multiplying by 5/9, i.e., $C = 5/9 \times (F - 32)$. This gives us a way of generating an arbitrary number of temperature scales by replacing 5/9 with a and 32 with b. This creates a new temperature unit related to Fahrenheit through the formula $a \cdot (F - b)$.

We can do the same with the Rasch model. Suppose that we know the abilities of the people are $\theta_1, \ldots, \theta_P$ on some unit of mathematics ability, "Unit A". We can create another unit of mathematics ability, "Unit B", by converting each θ_p to $\theta'_p = \theta_p - b$. Of course, when converting θ_p to θ'_p, we must also convert each β_i to $\beta'_i = \beta_i - b$ to correctly locate each item difficulty on the new scale. After doing this, we have

$$\log(O'_{pi}) = \theta'_p - \beta'_i = (\theta_p - b) - (\beta_i - b) = \theta_p - \beta_i = \log(O_{pi}),$$

so the change of scale does not change the probability of a correct response. This means that the two scales make identical behavioral predictions, so there is no way we can tell them apart.

We might also consider rescaling θ_p to some other unit of mathematics ability, "Unit C", by converting θ_p to $\theta''_p = (1/a) \cdot \theta_p$ and β_i to $\beta''_i = (1/a) \cdot \beta_i$. In this case, we have

$$\log(O''_{pi}) = \theta''_p - \beta''_i = (1/a) \cdot \theta_p - (1/a) \cdot \beta_i = (1/a) \cdot (\theta_p - \beta_i).$$

At first glance, it appears that the Rasch model distinguishes Unit A from Unit C, because all of the logits for Unit C are $1/a$ times their counterparts for Unit A. However, suppose we replace the logistic function $f(x)$ as the item response function with a logistic function whose argument is scaled by a, i.e., $f(a \cdot x)$. The logit of this function will be

$$a \cdot (\theta''_p - \beta''_i) = a \cdot [(1/a) \cdot (\theta_p - \beta_i)] = \theta_p - \beta_i.$$

This means that the probability corresponding to an ability measured in Unit A resulting from the logistic item response function will be the same as the probability corresponding to an ability measured in Unit C resulting from the scaled logistic item response function. Thus, there is no way to distinguish these situations using data.

Although items and persons can be considered to be measured on an interval scale (Fischer & Molenaar, 1995, Chapter 1), it follows from this discussion that their zero point and their scale are not determined. In practice, this means

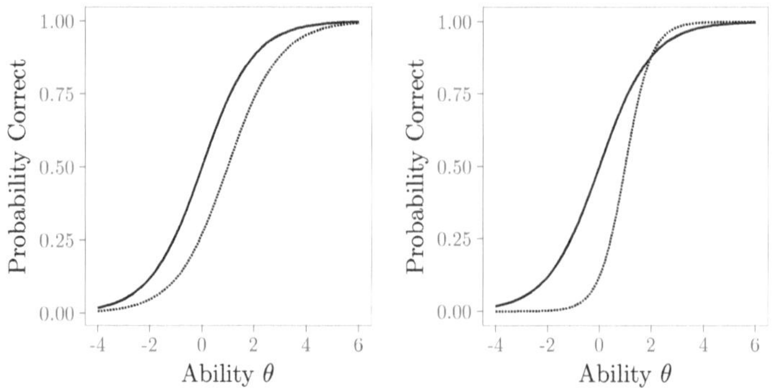

FIGURE 2.5: Figures used in the end-of-chapter exercises.

that we will need to select the zero point and scale of the item response function. This can be done by means of the following constraints. For selecting the zero point, there are at least three conventions: setting the difficulty of the first test item to be zero, forcing the difficulties to sum to zero, or forcing the abilities to sum to zero. The scale is typically set to 1 for the Rasch model.[9] This also corresponds to fixing the slope for the items. We will encounter these options again in the next chapter on parameter estimation.

2.5 Exercises

1. Figure 2.5 (left) shows the item characteristic curves for two items.
   (a) Read off the difficulty of the item with the solid ICC.
   (b) Which item is more difficult, the item with the solid ICC or the item with the dashed ICC? Why?
2. What is a sufficient statistic? What are the sufficient statistics for the parameters of the Rasch model?
3. Think of a situation where
   (a) the assumption of local stochastic independence of items is violated.
   (b) the assumption of local stochastic independence of persons is violated.
4. Do the item characteristic curves in Figure 2.5 (right) satisfy specific objectivity? How can you tell?

[9]Or to 1.7 to approximate the 1PNO, as already noted in Section 2.2.3.

5. You hear a researcher say that unfortunately the Rasch model does not hold for their new test. As a result, they have decided to use the sum score of each test taker for their further analyses, rather than estimating the test takers' abilities by means of the Rasch model. What are they missing?

3

Parameter Estimation

CONTENTS

To evaluate and apply the Rasch model in practical research, we need to be able to estimate its model parameters based on empirical data. Therefore, this chapter presents different approaches for estimating the person and item parameters of the Rasch model from observed data. Understanding this chapter requires a basic understanding of both maximum likelihood (often abbreviated as ML) and Bayesian estimation. Brief introductions to these topics can be found in Appendix B.1. Just like Chapter 2, this chapter is intentionally thorough.

In this chapter, we will present different approaches to estimating the parameters of the Rasch model from observed test data. All of these approaches can be used to estimate both the item and the person parameters, but do so in different ways. Two of the approaches – joint maximum likelihood (Section 3.1) and Bayesian inference (Section 3.4) – estimate the person and item parameters simultaneously. The other two approaches – conditional maximum likelihood (Section 3.2) and marginal maximum likelihood (Section 3.3) – estimate them separately, with the person parameters following in a second step (Section 3.5). Section 3.6 introduces the concepts of item and test information. From a practical perspective, the information is related to the uncertainty of the estimation. In Section 3.7 we discuss the sample size requirements for estimating the parameters of the Rasch model.

All of the approaches presented here rely on the likelihood function. As explained in Appendix B.1, the likelihood is the probability of the observed data, expressed as a function of the unknown model parameters. The likelihood contribution of person p's response to item i is (cf. Equation (2.3)):

$$L_{u_{pi}}(\theta_p, \beta_i) = \Pr(U_{pi} = u_{pi} \mid \theta_p, \beta_i) = \frac{\exp\{u_{pi} \cdot (\theta_p - \beta_i)\}}{1 + \exp(\theta_p - \beta_i)}.$$

We can compute the likelihood of person p's response to all of the test items $i = 1, \ldots, I$ by computing the product of the likelihoods of all of p's responses (see Section 2.4.2 for details), i.e.,

$$\begin{aligned} L_{\boldsymbol{u}_p}(\theta_p, \boldsymbol{\beta}) &= \prod_{i=1}^{I} \frac{\exp\{u_{pi} \cdot (\theta_p - \beta_i)\}}{1 + \exp(\theta_p - \beta_i)} \\ &= \frac{\exp(r_p \cdot \theta_p - \sum_{i=1}^{I} u_{pi} \cdot \beta_i)}{\prod_{i=1}^{I} [1 + \exp(\theta_p - \beta_i)]}. \end{aligned} \tag{3.1}$$

This is the starting point for all of the estimation approaches presented here. However, the way the person parameters are dealt with depends on the approach.

3.1 Joint Maximum Likelihood Estimation

In joint maximum likelihood (often abbreviated JML) estimation, we find the person and item parameters that maximize the joint likelihood in Equation (3.1). It makes sense to select the parameters that maximize the joint likelihood, because these parameters are the most likely to have generated the observed data.

In psychometrics (as well as in statistics in general), we typically try to collect samples that are as large as possible, because they provide the most information about the population we are interested in. Suppose that we want to estimate the average income of college graduates. It would be unwise to base our estimates on a sample containing just five people, since we know that the average of small samples can vary substantially. A random sample of college graduates will sometimes contain a millionaire or a pauper, even though neither will occur very often in a larger sample. These extreme values can have a large effect on the sample mean. As a consequence, we will not have much confidence in the accuracy of the sample mean for a small sample.

If we used a sample of size 100, the sample mean would be much less affected by the occasional outlier. Thus, we would have much more confidence in the accuracy of the sample mean for a sample of size 100 than a sample of size five. In statistical terms, this means that the variance of the sample mean estimator is smaller for a sample of size 100 than a sample of size five. Were we to use a sample of size 1000, the variance of the sample mean estimator would be even smaller than for the sample of size 100. The larger the sample we use, the closer to zero the variance will be. In fact, for an infinitely large sample, the variance will essentially be zero. Many estimators have this property, which is known as consistency.

While straightforward, JML estimation is rarely used, because it does not generally provide consistent estimators for the item parameters of a given test, even if the number of persons goes to infinity.[1] Further properties of JML estimation are discussed in De Ayala (2009) and Baker and Kim (2004).

In R, the joint likelihood can be numerically maximized using the `tam.jml()` function in the `TAM` package (Robitzsch, Kiefer, & Wu, 2021, cf. Table 9.1) as well as with general-purpose functions for estimating generalized linear models. However, since it is usually not recommended we do not provide example code.

3.2 Conditional Maximum Likelihood Estimation

One solution to the problem described above is to estimate the person and item parameters using a two-stage approach. In the first step, we condition on the sufficient statistics of the person parameters. This allows us to estimate the item parameters without knowing the person parameters. In the second step, the person parameters can be estimated using the item parameters estimated in the first stage. An important assumption of this procedure is that the item parameters are estimated with sufficient accuracy in the first step to use them in the second step.

As we have already noted, the likelihood of person p's responses to all of the test items $i = 1, \ldots, I$ is

$$L_{\boldsymbol{u}_p}(\theta_p, \boldsymbol{\beta}) = \frac{\exp(r_p \cdot \theta_p - \sum_{i=1}^{I} u_{pi} \cdot \beta_i)}{\prod_{i=1}^{I}[1 + \exp(\theta_p - \beta_i)]}.$$

This includes person p's test score, or row sum, r_p. Recall from Section 2.4.1 that a test taker's score is a sufficient statistic for their person parameter θ_p. This fact can be used to factorize the joint likelihood into two parts

$$L_{\boldsymbol{u}_p}(\theta_p, \boldsymbol{\beta}) = h(\boldsymbol{u}_p \mid r_p, \theta_p, \boldsymbol{\beta}) \cdot g(r_p \mid \theta_p, \boldsymbol{\beta}).$$

This works, because any joint probability can be split into the product of a conditional and a marginal probability. In this case, the joint probability is

[1] It should be noted that in general any maximum likelihood estimator is consistent (see an introductory text on mathematical statistics, like Casella and Berger (2002), for details). The lack of consistency of JML estimation is a result of the fixed test length for a given test: If we could also sample infinitely many test items (with the ratio of persons to items going to infinity, too), joint maximum likelihood for the item paremeters would be consistent (Molenaar, 1995). However, this is not a realistic scenario.

denoted by $\Pr(\boldsymbol{u_p}, r_p \mid \theta_p, \boldsymbol{\beta})$, the conditional probability by $h(\boldsymbol{u_p} \mid r_p, \theta_p, \boldsymbol{\beta})$ and the marginal probability by $g(r_p \mid \theta_p, \boldsymbol{\beta})$. The joint probability can be written as

$$\Pr(\boldsymbol{u_p}, r_p \mid \theta_p, \boldsymbol{\beta}) = \Pr(\boldsymbol{u_p} \mid \theta_p, \boldsymbol{\beta}) = L_{\boldsymbol{u}_p}(\theta_p, \boldsymbol{\beta}).$$

Here, we can omit the score r_p, because its information is already contained in the response pattern $\boldsymbol{u}_p$ of person p.

For us, the most interesting factor is the conditional likelihood $h(\boldsymbol{u}_p \mid r_p, \theta_p, \boldsymbol{\beta})$. It turns out that the person parameter cancels out in the conditional likelihood, so it only depends on p's test responses u_p, their marginal sums r_p and the item parameters $\boldsymbol{\beta}$. This allows us to estimate the item parameters independently from the person parameters.

We can compute the conditional likelihood by rearranging the factorization of the likelihood of person p's responses. This gives

$$h(\boldsymbol{u}_p \mid r_p, \theta_p, \boldsymbol{\beta}) = \frac{L_{\boldsymbol{u}_p}(\theta_p, \boldsymbol{\beta})}{g(r_p \mid \theta_p, \boldsymbol{\beta})}.$$

We already know $L_{\boldsymbol{u}_p}(\theta_p, \boldsymbol{\beta})$, so we only need to work out $g(r_p \mid \theta_p, \boldsymbol{\beta})$, the probability of observing a particular score r_p given p's ability and the difficulties of the items. The probability of a score r_p is the probability of observing a response pattern whose sum is r_p for a given ability θ_p. We can compute this probability by summing up the individual probabilities of all of the response patterns with r_p ones and $I - r_p$ zeros. Let Γ_{r_p} be the set of such response patterns. Then,

$$\begin{aligned} g(r_p \mid \theta_p, \boldsymbol{\beta}) &= \sum_{\boldsymbol{u}_p \in \Gamma_{r_p}} \Pr(\boldsymbol{u_p} \mid \theta_p, \boldsymbol{\beta}) \\ &= \sum_{\boldsymbol{u}_p \in \Gamma_{r_p}} \frac{\exp(r_p \cdot \theta_p - \sum_{i=1}^{I} u_{pi} \cdot \beta_i)}{\prod_{i=1}^{I}[1 + \exp(\theta_p - \beta_i)]}, \end{aligned}$$

after substituting the result from Equation (2.5) in Section 2.4.2.

We can simplify this expression using the fact that $\exp(x + y) = \exp(x) \cdot \exp(y)$ (see Appendix A). This allows us to factor the numerator into $\exp(r_p \cdot \theta_p) \cdot \exp(-\sum_{i=1}^{I} u_{pi}\beta_i)$, which gives

$$g(r_p \mid \theta_p, \boldsymbol{\beta}) = \sum_{\boldsymbol{u}_p \in \Gamma_{r_p}} \frac{\exp(r_p \cdot \theta_p) \cdot \exp(-\sum_{i=1}^{I} u_{pi} \cdot \beta_i)}{\prod_{i=1}^{I}[1 + \exp(\theta_p - \beta_i)]}.$$

Moreover, since r_p is constant for all elements of Γ_{r_p}, we can pull $\exp(r_p \cdot \theta_p)$ and $\prod_{i=1}^{I}[1 + \exp(\theta_p - \beta_i)]$ out of the sum, yielding

$$g(r_p \mid \theta_p, \boldsymbol{\beta}) = \frac{\exp(r_p \cdot \theta_p)}{\prod_{i=1}^{I}[1 + \exp(\theta_p - \beta_i)]} \cdot \sum_{\boldsymbol{u}_p \in \Gamma_p} \exp\left(-\sum_{i=1}^{I} u_{pi} \cdot \beta_i\right).$$

Item				
1	2	3	4	5
1	1	1	0	0
1	1	0	1	0
1	1	0	0	1
1	0	1	1	0
1	0	1	0	1
⋮	⋮	⋮	⋮	⋮

TABLE 3.1: Exemplary response patterns where three out of five items were answered correctly.

We illustrate how the sum over response patterns can be computed by example. Suppose that test taker p correctly answered three items of a five-item test correctly. Then, Γ_p is the set of response patterns with three correct items and two incorrect items. A few of these response patterns are listed in Table 3.1.

Let's look at the value of $\exp(-\sum_{i=1}^{I} u_{pi} \cdot \beta_i)$ for the first of these response patterns, $\boldsymbol{u}_p = (1, 1, 1, 0, 0)$. This is

$$\begin{aligned} \exp\left(-\sum_{i=1}^{I} u_{pi} \cdot \beta_i\right) &= \exp\{-(1 \cdot \beta_1 + 1 \cdot \beta_2 + 1 \cdot \beta_3 + 0 \cdot \beta_4 + 0 \cdot \beta_5)\} \\ &= \exp(-\beta_1 - \beta_2 - \beta_3) \\ &= e^{-\beta_1} \cdot e^{-\beta_2} \cdot e^{-\beta_3}. \end{aligned}$$

If we define $\varepsilon_i = e^{-\beta_i}$, then we can simplify $e^{-\beta_1} \cdot e^{-\beta_2} \cdot e^{-\beta_3}$ to $\varepsilon_1 \cdot \varepsilon_2 \cdot \varepsilon_3$. Similarly, the value for the second response pattern, $\boldsymbol{u}_p = (1, 1, 0, 1, 0)$, is

$$\exp\left(-\sum_{i=1}^{I} u_{pi} \cdot \beta_i\right) = e^{-\beta_1} \cdot e^{-\beta_2} \cdot e^{-\beta_4} = \varepsilon_1 \cdot \varepsilon_2 \cdot \varepsilon_4,$$

and so on. Thus, the sum of

$$\begin{aligned} \sum_{\boldsymbol{u}_p \in \Gamma_p} \exp\left(-\sum_{i=1}^{I} u_{pi} \cdot \beta_i\right) &= \varepsilon_1 \cdot \varepsilon_2 \cdot \varepsilon_3 + \varepsilon_1 \cdot \varepsilon_2 \cdot \varepsilon_4 + \varepsilon_1 \cdot \varepsilon_2 \cdot \varepsilon_5 \\ &\quad + \varepsilon_1 \cdot \varepsilon_3 \cdot \varepsilon_4 + \varepsilon_1 \cdot \varepsilon_3 \cdot \varepsilon_5 + \ldots \end{aligned}$$

This function is known in mathematics as an elementary symmetric function. The elementary symmetric functions are sums of products of a set of atoms. The number of terms in each product is the order of the elementary symmetric function. In our example, the atoms are $\varepsilon_1, \ldots, \varepsilon_5$. The order of the elementary symmetric function is three, because each product has three terms. Note that the order of the elementary symmetric function is equal to r_p. This is not by accident. Were we to repeat this exercise with $r_p = 2$ instead of three, we would get the elementary symmetric function of order two,

$$\begin{aligned}\gamma_2(\boldsymbol{\beta}) &= \varepsilon_1 \cdot \varepsilon_2 + \varepsilon_1 \cdot \varepsilon_3 + \varepsilon_1 \cdot \varepsilon_4 + \varepsilon_1 \cdot \varepsilon_5 + \varepsilon_2 \cdot \varepsilon_3 \\ &\quad + \varepsilon_2 \cdot \varepsilon_4 + \varepsilon_2 \cdot \varepsilon_5 + \varepsilon_3 \cdot \varepsilon_4 + \varepsilon_3 \cdot \varepsilon_5 + \varepsilon_4 \cdot \varepsilon_5.\end{aligned}$$

More generally, for a test with I items, we have I atoms $\varepsilon_1, \ldots, \varepsilon_I$, where $\varepsilon_i = e^{-\beta_i}$, resulting in the general notation

$$\sum_{\boldsymbol{u}_p \in \Gamma_p} \exp\left(-\sum_{i=1}^{I} u_{pi} \cdot \beta_i\right) = \gamma_{r_p}(\boldsymbol{\beta}).$$

This allows us to simplify the expression of $g(r_p \mid \theta_p, \boldsymbol{\beta})$ to

$$g(r_p \mid \theta_p, \boldsymbol{\beta}) = \frac{\exp(r_p \cdot \theta_p) \cdot \gamma_{r_p}(\boldsymbol{\beta})}{\prod_{i=1}^{I}[1 + \exp(\theta_p - \beta_i)]}.$$

Substituting for $g(r_p \mid \theta_p, \boldsymbol{\beta})$ in the definition of $h(\boldsymbol{u}_p \mid r_p, \theta_p, \boldsymbol{\beta})$ yields

$$\begin{aligned}h(\boldsymbol{u_p} \mid r_p, \theta_p, \boldsymbol{\beta}) &= \frac{L_{\boldsymbol{u_p}}(\theta_p, \boldsymbol{\beta})}{g(r_p \mid \theta_p, \boldsymbol{\beta})} \\ &= \frac{\left(\dfrac{\exp(r_p \cdot \theta_p - \sum_{i=1}^{I} u_{pi} \cdot \beta_i)}{\prod_{i=1}^{I}[1 + \exp(\theta_p - \beta_i)]}\right)}{\left(\dfrac{\exp(r_p \cdot \theta_p) \cdot \gamma_{r_p}(\boldsymbol{\beta})}{\prod_{i=1}^{I}[1 + \exp(\theta_p - \beta_i)]}\right)} \\ &= \frac{\exp(-\sum_{i=1}^{I} u_{pi} \cdot \beta_i)}{\gamma_{r_p}(\boldsymbol{\beta})},\end{aligned}$$

after dividing out the common $\exp(r_p \cdot \theta_p)$ and $\prod_{i=1}^{I}[1 + \exp(\theta_p - \beta_i)]$ terms. As promised, the fact that the marginal sum r_p is a sufficient statistic for θ_p allowed us to derive a conditional likelihood that does not depend on the person parameter θ_p. Thus, we can write $h(\boldsymbol{u}_p \mid r_p, \theta_p, \boldsymbol{\beta})$ as $h(\boldsymbol{u}_p \mid r_p, \boldsymbol{\beta})$, which no longer conditions on θ_p.

We can emphasize that $h(\boldsymbol{u}_p \mid r_p, \boldsymbol{\beta})$ is a likelihood by writing it as $L_{\boldsymbol{u}_p}(r_p, \boldsymbol{\beta})$. We can use the individual likelihoods of the people to work out the conditional likelihood for the entire data matrix. As discussed in Section 2.4.2, responses are assumed to be independent in the Rasch model. Thus,

$$\begin{aligned}L_{\boldsymbol{u}}(\boldsymbol{r}, \boldsymbol{\beta}) &= \prod_{p=1}^{P} \frac{\exp(-\sum_{i=1}^{I} u_{pi} \cdot \beta_i)}{\gamma_{r_p}(\boldsymbol{\beta})} \\ &= \frac{\exp(-\sum_{p=1}^{P} \sum_{i=1}^{I} u_{pi} \cdot \beta_i)}{\prod_{p=1}^{P} \gamma_{r_p}(\boldsymbol{\beta})},\end{aligned}$$

after pulling the product into the exponent. Switching the order of the sums we get that

$$L_{\boldsymbol{u}}(\boldsymbol{r}, \boldsymbol{\beta}) = \frac{\exp(-\sum_{i=1}^{I}\sum_{p=1}^{P} u_{pi} \cdot \beta_i)}{\prod_{p=1}^{P} \gamma_{r_p}(\boldsymbol{\beta})}.$$

The sums $\sum_{p=1}^{P} u_{pi}$ are simply the column sums c_i. This allows us to simplify the conditional likelihood as

$$L_{\boldsymbol{u}}(\boldsymbol{r}, \boldsymbol{\beta}) = \frac{\exp(-\sum_{i=1}^{I} c_i \cdot \beta_i)}{\prod_{p=1}^{P} \gamma_{r_p}(\boldsymbol{\beta})}. \tag{3.2}$$

The item parameters are estimated by finding the value of $\boldsymbol{\beta}$ that maximizes the conditional likelihood given $\boldsymbol{r}$. Owing to the complicated form of the conditional likelihood, we must apply numerical methods to determine the maximum likelihood estimate for $\boldsymbol{\beta}$. In R, we can use the `RM()` function in the `eRm` package (Mair, Hatzinger, & Maier, 2021). An example demonstrating the `RM()` function can be found in Chapter 6.

The lack of a unique origin for the latent scale (see Section 2.4.5 for details) forces us to use a linear constraint to identify the model. In practice, we usually use one of two possible constraints. The first possible constraint sets the value of the first item parameter to zero. In this case, we can interpret the β_i as the difficulty of item i relative to the difficulty of the first item. The second possible constraint sets the sum (and thus the average) of the β_i to be 0. In this case, we can interpret the value of β_i as the difficulty of item i relative to the average difficulty of all of the items. The scaling factor of the latent scale is defined implicitly by setting the slope of all items to be one.

Once the item parameters have been estimated, they can be substituted into the joint likelihood in order to estimate the person parameters, see Section 3.5.

3.3 Marginal Maximum Likelihood Estimation

Conditional maximum likelihood estimation replaces each test taker's person parameter with their corresponding sum score. This allows both the person and item parameters to be consistently estimated. Another approach to "get rid of" the person parameters when estimating the item parameters is marginal maximum likelihood estimation. In this approach, the person parameters are "averaged out" of the joint likelihood.

Marginal maximum likelihood estimation requires a marginal, or population, distribution for the person parameters. This distribution expresses the relative probability of each possible ability parameter. As De Ayala (2009) points out, marginal maximum likelihood estimation treats the abilities as random effects in the sense of mixed (or multilevel) models, whereas joint maximum likelihood estimation treats them as fixed effects.

A common choice of population distribution for the person parameters is the normal distribution. This makes sense when we can reasonably assume a symmetric distribution where the majority of people have person parameters near the mean, with just a few people having very high or very low values. Intelligence is a good example of this, since most people have average intelligence and just a few people have very high or very low intelligence.

When the population does not follow a normal distribution, assuming normality can be misleading. For example, when the population is heavily skewed, the normal distribution will do a poor job of capturing the ability distribution of the test takers. In this case, assuming normality can distort estimates of both the person and item parameters (Zwinderman & van den Wollenberg, 1990).

Let f denote the population distribution, so that $f(\theta_p)$ is the density of p's person parameter. The density $f(\theta_p)$ will be large when θ_p is common and small when θ_p is rare. To compute the marginal density of $\boldsymbol{u}_p$, we first compute the joint density of $\boldsymbol{u}_p$ and θ_p and then we marginalize θ_p from this joint density. By definition, the joint density is

$$\begin{aligned} f(\boldsymbol{u}_p, \theta_p \mid \boldsymbol{\beta}) &= \Pr(\boldsymbol{u}_p \mid \theta_p, \boldsymbol{\beta}) f(\theta_p) \\ &= L_{\boldsymbol{u}_p}(\theta_p, \boldsymbol{\beta}) f(\theta_p), \end{aligned} \tag{3.3}$$

since $L_{\boldsymbol{u}_p}(\theta_p, \boldsymbol{\beta}) = \Pr(\boldsymbol{u}_p \mid \theta_p, \boldsymbol{\beta})$.

We marginalize out θ_p by integrating $f(\boldsymbol{u}_p, \theta_p \mid \boldsymbol{\beta})$ with respect to θ_p. The resulting marginal likelihood of $\boldsymbol{u}_p$ is

$$L_{\boldsymbol{u}_p}(\boldsymbol{\beta}) = \int_{-\infty}^{\infty} f(\boldsymbol{u}_p, \theta_p \mid \boldsymbol{\beta}) d\theta_p.$$

Often, the marginal likelihood is written in terms of the likelihood and population distribution, rather than the joint density. By combining the previous two equations, we obtain the marginal likelihood of $\boldsymbol{u}_p$

$$L_{\boldsymbol{u}_p}(\boldsymbol{\beta}) = \int_{-\infty}^{\infty} L_{\boldsymbol{u}_p}(\theta_p, \boldsymbol{\beta}) f(\theta_p) d\theta_p.$$

The full likelihood is obtained by multiplying the likelihoods for the individual response patterns, resulting in

$$L_{\boldsymbol{u}}(\boldsymbol{\beta}) = \prod_{p=1}^{P} L_{\boldsymbol{u}_p}(\boldsymbol{\beta}) = \prod_{p=1}^{P} \int_{-\infty}^{\infty} L_{\boldsymbol{u}_p}(\theta_p, \boldsymbol{\beta}) f(\theta_p) d\theta_p.$$

The marginal likelihood can be understood mathematically in the following way. Using Equation (3.3), the marginal likelihood of $\boldsymbol{u}_p$ can be expressed as

$$L_{\boldsymbol{u}_p}(\boldsymbol{\beta}) = \int_{-\infty}^{\infty} L_{\boldsymbol{u}_p}(\theta_p, \boldsymbol{\beta}) f(\theta_p) d\theta_p = E_{\theta_p}[L_{\boldsymbol{u}_p}(\theta_p, \boldsymbol{\beta})],$$

where $E_{\theta_p}[L_{\boldsymbol{u}_p}(\theta_p, \boldsymbol{\beta})]$ is the expected value of the joint likelihood over $f(\theta_p)$. This tells us that the contribution of person p's responses to the marginal likelihood of $\boldsymbol{\beta}$ is the average likelihood of his or her responses over the population. Thus, marginal maximum likelihood estimation finds the value of $\boldsymbol{\beta}$ that maximizes the overall probability of the observed response patterns while considering the assumed ability distribution in the population.

In R, marginal maximum likelihood estimation is provided by both the `mirt` (Chalmers, 2021) and `TAM` packages (Robitzsch et al., 2021). We demonstrate how to use these packages in Chapter 7 and Chapter 8, respectively. There is also an older package using marginal maximum likelihood estimation, `ltm` (Rizopoulos, 2006), but it is no longer maintained. Thus, we no longer recommend using it.

Just like conditional maximum likelihood, the marginal maximum likelihood approach requires that the person parameters are estimated in a second step, often called scoring, see Section 3.5.

3.4 Bayesian Estimation

Bayesian estimation is an increasingly popular way of estimating the parameters of the Rasch model (Fox, 2010). Like joint maximum likelihood, Bayesian estimation simultaneously estimates both the person and item parameters. However, while joint maximum likelihood estimation finds the values of $\boldsymbol{\theta}$ and $\boldsymbol{\beta}$ by maximizing the joint likelihood, Bayesian estimation uses Bayes' rule to find the posterior density, $f(\boldsymbol{\theta}, \boldsymbol{\beta} \mid \boldsymbol{u})$. For a primer on Bayesian inference, see the Appendix B.1.4.

For the Rasch model, Bayes' rule states that

$$f(\boldsymbol{\theta}, \boldsymbol{\beta} \mid \boldsymbol{u}) = \frac{\Pr(\boldsymbol{u} \mid \boldsymbol{\theta}, \boldsymbol{\beta}) f(\boldsymbol{\theta}, \boldsymbol{\beta})}{\Pr(\boldsymbol{u})}. \tag{3.4}$$

The first term in the numerator, $\Pr(\boldsymbol{u} \mid \boldsymbol{\theta}, \boldsymbol{\beta})$, is the joint likelihood that we discussed in Section 3.1.[2] The second is the joint prior distribution for $\boldsymbol{\theta}$ and $\boldsymbol{\beta}$. The denominator is the average probability of the observed data over the joint prior distribution.

The use of prior distributions is a key distinction between the Bayesian approach and the "frequentist" approaches presented in the previous sections. Though, theoretically, we are free to choose the prior distribution to be whatever we want, it is typically chosen to reflect reasonable assumptions about the locations of the item parameters.

[2]Here we denote the likelihood as a probability (with Pr instead of L), because this makes it easier to recognize Bayes' rule as it is presented in the appendix.

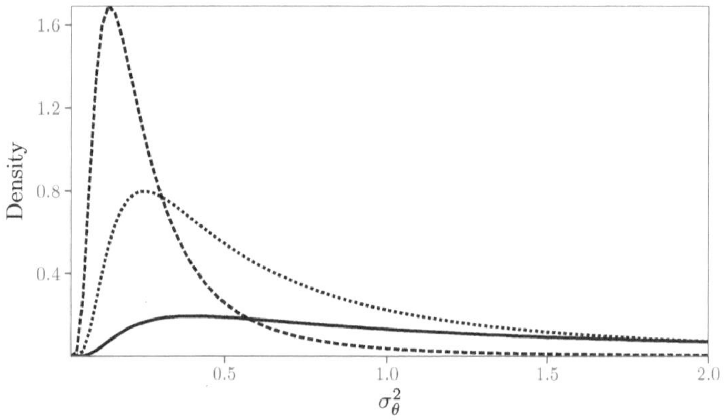

FIGURE 3.1: The inverse-χ^2 distribution of σ^2_θ with degrees of freedom ν_θ equal to 0.5 (solid line), 2 (dotted line) and 3 (dashed line).

For the Rasch model, we typically assume that the person parameters are independent draws from a normal distribution with a mean of zero and a variance of σ^2_θ. Note that here we are using θ as a subscript to indicate the model parameter that each distributional parameter is associated with. We choose a normal distribution for the reasons discussed in Section 3.3 for marginal maximum likelihood estimation. We set the mean to zero in order to fix the location of the latent scale. This is often abbreviated as $\theta_p \sim N(0, \sigma^2_\theta)$, using the distributional notation introduced in Section 2.3.2. In this expression, $N(0, \sigma^2_\theta)$ denotes a normal distribution whose mean is zero and whose variance is σ^2_θ.

Rather than assume a fixed value for σ^2_θ, we would like to infer it from the observed test data. To do this, we employ an inverse-χ^2 prior distribution for σ^2_θ. The inverse-χ^2 is the distribution of the random variable $1/Z$, when Z has a χ^2 distribution. The χ^2 distribution is a common distribution in statistics, because it is the sampling distribution of the test statistic for a number of common statistical tests, such as the likelihood ratio test. The inverse-χ^2 distribution is a common prior for variance parameters, because it only has non-zero density for positive values and has useful computational properties. The inverse-χ^2 distribution is shown for a number of different degrees of freedom ν_θ in Figure 3.1. The most common choice of ν_θ is 0.5, which is represented by the solid line. We will denote that σ^2_θ has an inverse-χ^2 distribution with ν_θ degrees of freedom by writing $\sigma^2_\theta \sim \text{Inv-}\chi^2(\nu_\theta)$.

The item parameters are also typically assumed to be independent draws from a normal distribution. The mean of this distribution is μ_β and its variance is σ^2_β. This is often abbreviated as $\beta_i \sim N(\mu_\beta, \sigma^2_\beta)$. We allow both the mean μ_β and the variance σ^2_β to be free parameters, and infer their values from the data. We use an improper uniform prior density for μ_β. The improper uniform density assigns equal density to every possible value of μ_β, i.e., the density of μ_β is 1 everywhere. It is called improper, because it is not a true probability distribution (its integral does not exist). We will write this as $f(\mu_\beta) \propto 1$. We again use an inverse-χ^2 distribution with ν_β degrees of freedom for σ^2_β, which we will write $\sigma^2_\beta \sim \text{Inv-}\chi^2(\nu_\beta)$.

Given these definitions, the joint prior distribution is

$$f(\boldsymbol{\theta}, \boldsymbol{\beta}, \sigma^2_\theta, \mu_\beta, \sigma^2_\beta \mid \mu_\beta, \sigma^2_\beta) = \prod_{p=1}^{P} \phi(\theta_p \mid 0, \sigma^2_\theta) \cdot \prod_{i=1}^{I} \phi(\beta_i \mid \mu_\beta, \sigma^2_\beta) \cdot f_{1/\chi^2}(\sigma^2_\theta \mid \nu_\theta) \cdot f_{1/\chi^2}(\sigma^2_\beta \mid \nu_\beta),$$

where $\phi(\cdot \mid \mu, \sigma^2)$ is the probability density function of a normal distribution whose mean is μ and whose variance is σ^2 and $f_{1/\chi^2}(\cdot \mid \nu)$ denotes the probability density function of an inverse-χ^2 distribution with ν degrees of freedom. We did not include a prior term for μ_β, because its density is always 1. This is often written as

$$\begin{aligned}
\theta_p \mid \sigma^2_\theta &\sim N(0, \sigma^2_\theta) \\
\beta_i \mid \mu_\beta, \sigma^2_\beta &\sim N(\mu_\beta, \sigma^2_\beta) \\
f(\mu_\beta) &\propto 1 \\
\sigma^2_\theta &\sim \text{Inv-}\chi^2(\nu_\theta) \\
\sigma^2_\beta &\sim \text{Inv-}\chi^2(\nu_\beta).
\end{aligned}$$

The joint posterior $f(\boldsymbol{\theta}, \boldsymbol{\beta}, \sigma^2_\theta, \mu_\beta, \sigma^2_\beta \mid \boldsymbol{u}, \nu_\theta, \nu_\beta)$ is defined by Equation (3.4). We can compute the numerator by substituting the joint likelihood for $\Pr(\boldsymbol{u} \mid \boldsymbol{\theta}, \boldsymbol{\beta})$ and $f(\boldsymbol{\theta}, \boldsymbol{\beta}, \sigma^2_\theta, \mu_\beta, \sigma^2_\beta \mid \nu_\theta, \nu_\beta)$ for the prior. The denominator, $\Pr(\boldsymbol{u} \mid \nu_\theta, \nu_\beta)$, cannot be computed analytically. To deal with this, we sample the joint posterior using Markov chain Monte Carlo (MCMC) methods. These methods provide a way to sample probability distributions that are only defined up to a proportionality constant. In Bayesian inference, that proportionality constant is the unknown denominator in Bayes' rule.

For the majority of applications, we do not need to deal with the details of implementing MCMC methods, as a number of software packages exist which automate this process. The best known of these packages are WinBUGS (Lunn, Thomas, Best, & Spiegelhalter, 2000), OpenBUGS (Lunn, Spiegelhalter, Thomas, & Best, 2009), JAGS (Plummer, 2017) and Stan (Stan Development Team, 2021). Each of these provides a way to specify the prior, likelihood and data, which define the numerator in Bayes' rule. From there, each employs an MCMC algorithm to draw samples from the posterior distribution. In the second part of the book, we demonstrate how Stan can be used to sample the posterior of the Rasch model.

We can use samples from the posterior distribution to produce point and interval estimates of the parameters of interest. Bayesian analyses typically use the posterior mean as a point estimate of the unknown parameters. The posterior mean can be computed from posterior samples by computing the sample mean of each unknown parameter. Suppose that we have $S = 1000$ posterior samples and let $\theta_p^{(s)}$ and $\beta_i^{(s)}$ be the s-th samples of θ_p and β_i. Then, the posterior mean estimates of θ_p and β_i are

$$\hat{\theta}_p = \frac{1}{S}\sum_{s=1}^{S}\theta_p^{(s)} \quad \text{and} \quad \hat{\beta}_i = \frac{1}{S}\sum_{s=1}^{S}\beta_i^{(s)}.$$

We can compute $(1-\alpha) \times 100\%$ highest posterior density intervals by empirically computing the appropriate quantiles. In the case of a 95% highest posterior density interval, these are the 0.025 and 0.975 quantiles.

3.5 Person Parameter Estimation

Two of the approaches presented above are frameworks that allow the estimation of the item and person parameters at the same time: joint maximum likelihood and Bayesian estimation. The other two approaches – conditional and marginal maximum likelihood estimation – first estimate the item parameters. These item parameter estimates can then be used for estimating the person parameters.

After estimating the item parameters using conditional maximum likelihood estimation, we can plug the estimated values into the joint likelihood and apply maximum likelihood to estimate the person parameters. In doing so, the uncertainty from estimating the item parameters is typically ignored. This can lead to confidence intervals for the person parameters that are too narrow (Cheng & Yuan, 2010; Tsutakawa & Johnson, 1990). We will encounter a similar problem later when thinking about how to quantify goodness-of-fit in Section 4.3.4.

Another drawback of the conditional maximum likelihood approach is its inability to estimate the ability of test takers who correctly answered all or none of the test items correctly.[3] Mathematically, the maximum likelihood estimate tends to plus or minus infinity for these test takers (Hoijtink & Boomsma, 1995). Intuitively, this makes sense. The only thing we know, for example, about a person correctly answering every item is that their ability is at least as large as the difficulty of the hardest item. We need a test taker to incorrectly answer an item to pin down their ability more exactly. Therefore, it is reasonable not to provide an estimate for a test taker with a perfect score (Mair & Hatzinger, 2007).

The different R packages handle this issue in different ways. The `eRm` package uses a spline-based approach, where it is possible to extrapolate values

[3]Note that, in principle, the same problem applies to items that were solved by all or none of the test takers, but usually this can be avoided by using a large, diverse sample.

for these test takers using estimates from all of the other test takers (Mair et al., 2021). This will produce estimates, but these estimates will be affected by the quality of the extrapolation. We will illustrate in the second part of the book in Section 6.4 that the `person.parameter` function in `eRm` determines the person parameters in this way.

In principle, the same strategy of plugging the item parameter estimates into the joint likelihood is possible after marginal maximum likelihood estimation. This approach is available as option `ML` and `MLE` in the `mirt` and `TAM` packages, respectively (cf. Table 9.1). The `mirt` package will actually return a person parameter estimate of `Inf` or `-Inf` for plus or minus infinity for test takers who answered all or none of the items correctly. In `TAM`, extreme raw scores are adjusted by a small value – that is, perfect scores are slightly reduced and null scores are slightly increased – to allow the application of ML estimation.

Another drawback of ML estimation of the person parameters is that it can be inaccurate when the number of items is small. Its bias becomes negligible when the test length goes to infinity, but can be problematic in short tests with limited item spread, particularly for extreme values of θ, as illustrated by Hoijtink and Boomsma (1995). A popular method for dealing with this problem is to use Warm's (1989) weighted likelihood estimator (WLE, as option `WLE` in Table 9.1).

The WLE is mathematically similar to the maximum a posteriori estimator (see below), though it is derived differently. The WL estimator $\hat{\boldsymbol{\beta}}$, when applied to a person p, assigns a weight $w(\theta_p)$ to each possible value of θ_p and maximizes the weighted likelihood

$$w(\theta_p) \cdot L_{\boldsymbol{u}_p}(\theta_p; \boldsymbol{\beta} = \hat{\boldsymbol{\beta}}).$$

By introducing these weights, the resulting estimator can compensate the bias of the ML estimator. The weights are chosen suitably so that the bias of the WLE becomes 0. The WLE is therefore constructed to have better performance for most test takers than the ML estimator, particularly on shorter tests (Warm, 1989). For the Rasch and 2PL models, the WLE is equivalent to the Bayesian maximum a posteriori estimator (see below) using a type of uninformative prior distribution known as a Jeffreys' prior as the prior distribution for the ability parameter (Magis & Raîche, 2012a). Both the weighted likelihood and Bayesian approaches provide estimates even for test takers who answered all or none of the items correctly.

Bayesian estimators, such as the maximum a posteriori (MAP) and expected a posteriori (EAP) estimators, are another alternative for estimating the person parameters in a second step after the item parameters. The EAP estimator is the expected value of the posterior distribution of θ_p given the test taker's responses $\boldsymbol{u}_p$ and the item parameter estimates $\hat{\boldsymbol{\beta}}$ from the first step. We can compute the posterior distribution by applying Bayes' rule

$$f(\theta_p \mid \boldsymbol{u}_p, \boldsymbol{\beta} = \hat{\boldsymbol{\beta}}) = \frac{L_{\boldsymbol{u}_p}(\theta_p, \boldsymbol{\beta} = \hat{\boldsymbol{\beta}}) \cdot f(\theta_p)}{L_{\boldsymbol{u}_p}(\boldsymbol{\beta} = \hat{\boldsymbol{\beta}})},$$

where $f(\theta_p)$ is the prior distribution (typically we assume a normal distribution for the person parameters, see also Section 3.3 and Section 3.4) and $L_{\boldsymbol{u}_p}(\theta_p, \boldsymbol{\beta} = \hat{\boldsymbol{\beta}})$ is the likelihood for the Rasch model given the observed response, item parameters, and the person parameter θ_p. The denominator marginalizes θ_p from $L_{\boldsymbol{u}_p}(\theta_p, \boldsymbol{\beta} = \hat{\boldsymbol{\beta}})$ to obtain the overall likelihood of p's responses given the item parameters are $\hat{\boldsymbol{\beta}}$.

The MAP estimator is similar to the EAP estimator, except that it uses the mode of the posterior distribution of θ_p rather than the mean.

A selection of these approaches is offered in the different R packages presented in the second part of this book (cf. Table 9.1). Overall, the different estimation techniques tend to give similar ability estimates, but often diverge at the extremes, as illustrated in Section 7.4. As outlined above, this can be due to differences in the behavior of the different person parameter estimation approaches.

That said, the person parameter estimation in the second step also relies on the results of the item parameter estimation in the first step. For example, marginal maximum likelihood assumes normality and can give inaccurate estimates of the item parameters when the true distribution of the person parameters is non-normal, e.g., highly skewed. This in turn can affect the estimation of the person parameters. A similar phenomenon can occur in Bayesian estimation when the prior distribution deviates strongly from the true parameter distribution.

3.6 Item and Test Information

The Rasch ICC shown in Figure 3.2 illustrates that the slope of an ICC is steepest in the center, i.e., where the ability is equal to the difficulty of the item. The larger the slope of the ICC, the larger the difference in probability for a given difference in ability. The larger differences in probability allow us to more accurately estimate the ability of a test taker when it is near the difficulty of the item than when it is not. This principle, which is formally captured by the information of an item, is also used in computerized adaptive testing (cf. Section 12.1).

This also means that we will be more certain about person parameter estimates near the difficulty of the item. Our certainty about a person parameter estimate is indicated by the width of its confidence interval for maximum likelihood estimators or highest posterior density (HPD) interval for the Bayesian

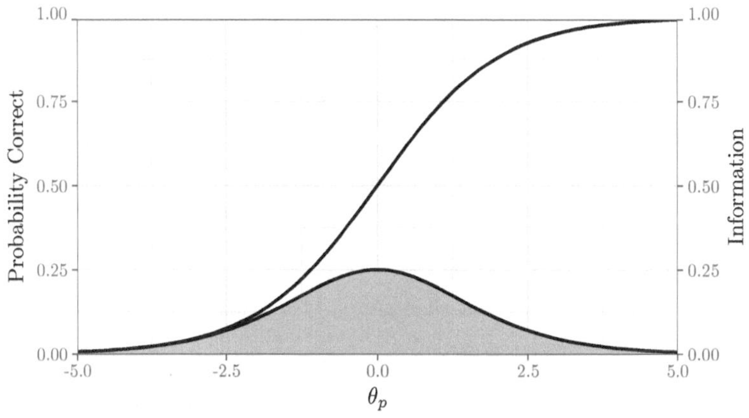

FIGURE 3.2: Item characteristic curve (top) and item information (bottom).

estimator. The confidence interval for the estimated person parameter is constructed in a way that ensures it will cover the true value of the person parameter 95% of the time. The HPD interval is the shortest interval containing 95% of the posterior density. In order to cover the true value 100% of the time, the confidence interval or HPD would need to be infinitely wide. 95% represents a good compromise, in the same way that the 0.05 significance level represents a good compromise for statistical tests.

Mathematically, the so-called Fisher information describes the amount of information the responses to a test provide at any position on the person parameter continuum. For the Rasch model, it simplifies to (Fischer & Molenaar, 1995, p.55)

$$\begin{aligned}\mathcal{I}(\theta_p) &= -E\left(\frac{\partial^2 \log \Pr(\boldsymbol{U}_p = \boldsymbol{u}_p \mid \theta_p, \boldsymbol{\beta})}{\partial^2 \theta_p}\right) \\ &= \sum_{i=1}^{I} \Pr(U_{pi} = 1 \mid \theta_p, \beta_i) \cdot \Pr(U_{pi} = 0 \mid \theta_p, \beta_i). \end{aligned} \tag{3.5}$$

In Equation (3.5) we see that the test information $\mathcal{I}(\theta_p)$ for the complete set of items forms a sum over the components for the individual items. This means that the information is additive over the items and a single item i provides the following item information $\mathcal{I}_i(\theta_p)$ (Fischer & Molenaar, 1995; Baker & Kim, 2004; De Ayala, 2009):

$$\mathcal{I}_i(\theta_p) = \Pr(U_{pi} = 1 \mid \theta_p, \beta_i) \cdot (1 - \Pr(U_{pi} = 1 \mid \theta_p, \beta_i)) \tag{3.6}$$

For the Rasch model, the information is equivalent to the first derivative, i.e., the slope, of an item's ICC (but note that this is not generally the case).[4]

As we have already argued intuitively above, we can see from Equation 3.6 and Figure 3.2 that the information is highest in the center, i.e., at the location of the item difficulty. There, both the probability of solving and of not solving the item are 0.5, giving 0.25 when multiplied. For any other location, the product of the two probabilities is lower. For example, when the probability of solving the item is 0.2 and of not solving it 0.8 – or vice versa – their product is 0.16.

We can see in Figure 3.2 that the shape of the item information is symmetric and roughly bell-shaped. Its form resembles that of a normal distribution. Remember that we have already discussed in Section 2.2.3 that the cumulative distribution function of the standard normal distribution, that is used in the 1PNO model, has a shape that is very similar to that of the logistic ICC of the Rasch model. Accordingly, the derivative of the logistic ICC looks very similar to a normal distribution. Its maximum value is at the location of the item's difficulty, and it is highest in the vicinity of this point.

We have also seen in Equation 3.5 that the information of the complete test is the sum over the information that the individual items provide at a certain location. Mathematically, this formula follows from the independence of the individual responses. Practically speaking, this means that a test provides most information in areas of the ability continuum where many items are located. This will be further illustrated in the second part of the book with examples in R. It also tells us that for comparable item locations a longer test will carry more information than a shorter test.

Many tests contain primarily items of moderate difficulty and only few very easy and very difficult items. It follows that they will provide more information about test takers of average ability than about test takers with very high or very low ability.

Under weak technical conditions, maximum likelihood estimators are asymptotically normally distributed (e.g., Casella & Berger, 2002). The corresponding confidence interval for the person parameter estimate $\hat{\theta}_p$ is

$$\left[\hat{\theta}_p \pm z_{1-\alpha/2} \cdot \mathcal{I}(\hat{\theta}_p)^{-1/2}\right].$$

[4]This can be shown using the differentiation rules for quotients and exponentials as well as the chain rule from Appendix A:

$$\frac{\partial \Pr(U_{pi}=1 \mid \theta_p, \beta_i)}{\partial \theta_p} = \frac{e^{(\theta_p-\beta_i)} \cdot 1 \cdot \left(1+e^{(\theta_p-\beta_i)}\right) - e^{(\theta_p-\beta_i)} \cdot e^{(\theta_p-\beta_i)} \cdot 1}{\left(1+e^{(\theta_p-\beta_i)}\right)^2} =$$

$$\frac{e^{(\theta_p-\beta_i)} + \left(e^{(\theta_p-\beta_i)}\right)^2 - \left(e^{(\theta_p-\beta_i)}\right)^2}{\left(1+e^{(\theta_p-\beta_i)}\right)^2} = \frac{e^{(\theta_p-\beta_i)}}{\left(1+e^{(\theta_p-\beta_i)}\right)^2} = \frac{e^{(\theta_p-\beta_i)}}{1+e^{(\theta_p-\beta_i)}} \cdot \frac{1}{1+e^{(\theta_p-\beta_i)}} =$$

$$\Pr(U_{pi}=1 \mid \theta_p, \beta_i) \cdot (1-\Pr(U_{pi}=1 \mid \theta_p, \beta_i)) = \mathcal{I}_i(\theta_p).$$

Here $\mathcal{I}(\hat{\theta}_p)$ is the test information at $\hat{\theta}_p$. We can see from the expression for the confidence interval that $\mathcal{I}(\hat{\theta}_p)^{-1/2}$ is the standard deviation of $\hat{\theta}_p$ (which for estimates is also termed standard error) and $z_{1-\alpha/2}$ determines the number of standard deviations needed to ensure the desired coverage. For example, when we want a 95% confidence interval, we set $\alpha = 0.05$ resulting in $1-\alpha/2 = 0.975$ for each of the two sides. Then, $z_{1-\alpha/2} = z_{0.975}$, the 97.5% quantile of the standard normal distribution, which is 1.96. This factor determines the width of the confidence interval.[5]

The Bayesian posterior distribution is also influenced by the information in a test. In this case, the posterior distribution of the average test taker will have a smaller variance than the posterior distribution of an exceptional test taker. This leads to shorter HPD intervals for typical test takers by comparison to extreme test takers. For very large samples, the HPD interval for the person parameter will be similar to the confidence interval for the person parameter. However, for most tests, the two intervals will differ as a result of the prior distribution.

We can also switch the roles of items and test takers and define the information given by a person sample for the estimation of an item parameter. Since items and test takers have analogous roles in the Rasch model, the main results that we obtained above also hold under this perspective. As a first important result, we get that the information given by a sample is simply the sum of the information given by the individual test takers. Conceptually, this follows because the responses of the individual respondents can be considered as independent under the Rasch model. A second important result is that the information provided by a single test taker p on item i increases when the test taker's ability is close to the item's difficulty.

As a consequence, we find that the estimation of an item parameter β_i is more precise in samples where many test takers have ability parameters close to β_i. We will illustrate this later.

3.7 Sample Size Requirements

The estimation of item parameters based on an observed sample of responses is often termed the calibration of the items. In general, a larger calibration sample allows a more accurate estimation of the item parameters, although other factors affect the accuracy of the estimation as well. For instance, the difficulty of an item can be estimated more accurately if the item is neither too easy nor too difficult for the sample of test takers. Therefore, factors that influence the estimation accuracy include the alignment and shape of the item

[5]As already noted in Section 3.2, the item parameters are usually treated as fixed here. This can lead to confidence intervals for the person parameters that are too short, especially for extreme abilities (Cheng & Yuan, 2010).

and person parameter distributions, the number of items, and the estimation technique.

Several publications have addressed the question which sample size is typically required for working with the Rasch model and how it is affected by these and other factors. For instance, De Ayala (2009) gives the rough guideline that a calibration sample should contain at least several hundred respondents and, among other references, mentions an earlier paper from Wright (1977) stating that a calibration sample of 500 would be more than adequate. Later De Ayala (2009) also suggests 250 or more respondents for fitting a Partial Credit model. Since the Partial Credit model is a generalization of the Rasch model with more item parameters (see Section 11.1), this implies that the suggested sample size of 250 should also suffice for fitting a Rasch model. More recent studies have investigated the application of the Rasch model with sample sizes as small as 100 respondents (e.g., Steinfeld & Robitzsch, 2021; Suárez-Falcón & Glas, 2003). We agree with De Ayala (2009) that such guidelines should not be interpreted as hard-and-fast rules, but that an adequate sample size depends on the conditions and goals of the analysis.

A more elaborate method for determining the necessary sample size is power analysis. Here, the desired estimation accuracy or risk of false-positive and false-negative statistical test results need to be formalized before the analysis. The necessary sample size is then determined based on these considerations. Publications that address power analysis specifically for the Rasch model are Draxler (2010) and Draxler and Alexandrowicz (2015). Users of R can also carry out simulation studies such as those reported in the literature themselves to replicate and extend the results reported there (see Mair, 2018, Section 4.5, for exemplary R code for the 2PL model, cf. Chapter 4, using the `mirt` package).

As we will also see in the second part of the book, an indication that the employed sample size was not sufficient for reliably estimating the model parameters are large standard errors or wide confidence intervals. In that case, the estimates are not trustworthy enough to report and interpret.

3.8 Exercises

1. Figure 3.3 (left) shows the likelihood for a range of values for the person parameter θ_p when the item parameters are known. Determine the maximum likelihood estimate of θ_p from the graph.
2. (a) Compare joint maximum likelihood to conditional maximum likelihood.

   (b) Compare conditional maximum likelihood to marginal maximum likelihood.

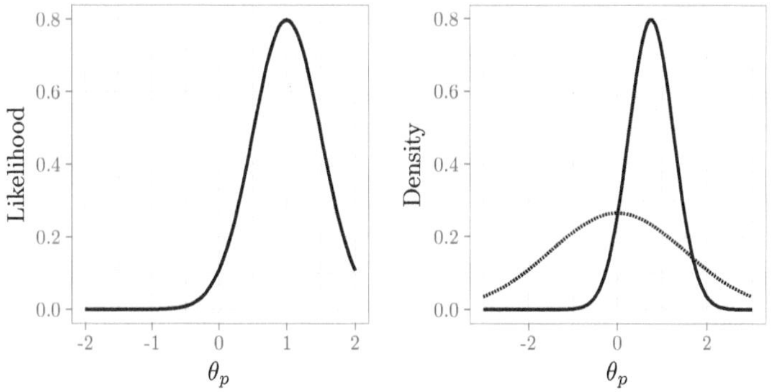

FIGURE 3.3: Figures for the end of chapter exercises. Likelihood, prior and posterior distributions for θ_p for known item parameters.

3. Figure 3.3 (right) shows the prior and posterior for a range of values for the person parameter θ_p when the item parameters are known. Determine which curve is the prior and which is the posterior.

4. Suppose you wanted to use the posterior median to estimate $\boldsymbol{\theta}$ and $\boldsymbol{\beta}$, rather than the posterior mean. How could you do this using MCMC samples?

5. In this advanced exercise – which requires a little more enthusiasm for mathematical brainteasers as well as a little more time to solve – we derive an additional estimation method that allows us to calculate the item parameters of the Rasch model by hand (or by calculator).

    - Consider two items 1 and 2 under the Rasch model, with item difficulty parameters β_1 and β_2. Consider a person with ability parameter θ_1. What is the probability that this person solves item 1, but not item 2? What is in turn the probability that this person solves item 2, but not item 1? What is the ratio of these two probabilities?
    - We now define two new terms $\epsilon_1 = \exp(-\beta_1)$ and $\epsilon_2 = \exp(-\beta_2)$. How can the ratio calculated in the previous step be expressed using ϵ_1 and ϵ_2? Does this ratio depend on the ability parameter?
    - To estimate the ϵ parameters (and thus the item difficulty parameters β), we now set the restriction that $\prod_i \epsilon_i = 1$. Show that this is equivalent to $\sum_i \beta_i = 0$.
    - Suppose we have a sufficiently large sample of test takers working on items 1 and 2, and some other items. Let p_{12} denote the

number of respondents that solve item 1, but not item 2. Analogously, let p_{21} denote the number of respondents that solve item 2, but not item 1. Can we estimate $\frac{\epsilon_1}{\epsilon_2}$ based on p_{12} and p_{21}?

- Given the relationship found in the last step and using $\prod_i \epsilon_i = 1$, how can we estimate ϵ_i for an arbitrary test item?

In the literature, this method is called the explicit procedure for item parameter estimation (Fischer & Scheiblechner, 1970).

4

Test Evaluation

CONTENTS

So far, in Chapter 2 we have introduced the theoretical properties or assumptions of the Rasch model and in Chapter 3 we have discussed how its parameters can be estimated from item response data. In this chapter, we want to address how we can check empirically whether the Rasch model is appropriate for data from a certain psychological test.

At this point it makes a big difference philosophically whether we formulate this question from a measurement or modeling point of view (cf. Section 1.2): Do we want to find out if "the data fit the model" or if "the model fits the data"? From a measurement point of view, remember that "the statistical model is being used as a means of quality control of the items" (Bond & Fox, 2007) and items that do not pass this quality control need to be revised or excluded from the test. From a modeling point of view, "the statistical model is used to describe the items" (Bond & Fox, 2007), and it might be necessary to move on to a model with more parameters if the Rasch model is too strict to adequately describe the data.

This chapter introduces a potpourri of the most commonly used graphical checks, descriptive indices, and statistical tests from different traditions that are available in R. The properties of statistical tests for the Rasch model have been evaluated and compared, for instance, by Maydeu-Olivares and Montaño (2013) and Debelak (2019). Most of the tests evaluated in these studies have been implemented in R and are presented in the following. The practical application of the approaches covered in this chapter is addressed in the second part of the book.

Section 4.1 introduces three means of graphical assessment for the Rasch model. Section 4.2 introduces formal statistical tests for the invariance of item and person parameters. Section 4.3 discusses a variety of commonly used tests and indices for checking the goodness-of-fit of the Rasch model, some of which can also be applied to other IRT models. Section 4.4 discusses separation indices that have been suggested in the Rasch framework and can be used to estimate the reliability of a psychological test. Section 4.5 introduces IRT models for dichotomous items that have more item parameters than the Rasch model, and addresses approaches for comparing the fit of different models as well as sample size considerations.

Note that most of the statistical tests and indices presented here will be tailored specifically to the Rasch model, and even for those which are applicable to more general models, we have simplified the notation according to the Rasch model. Understanding this chapter thus only requires basic knowledge of statistical tests, and the necessary background material is presented in Appendix B.2. That said, the principles that underlie Rasch model checks are also applied in many other statistical tests. The interested reader can learn more about the general principles of likelihood ratio tests, Wald tests, χ^2 goodness-of-fit tests, etc. in advanced statistics textbooks (e.g., Casella & Berger, 2002).

4.1 Graphical Assessment

The first means for test evaluation we would like to present are graphical methods. The first one, the person item map, shows whether the person sample covers the spread of the items and vice versa. The second approach of comparing expected and empirical ICCs can help detect misfitting items. The third one, the graphical test, is a visual test for DIF.

4.1.1 Person Item Map

We have learned in the previous chapters that the Rasch model places persons and items on the same latent scale, and that the accuracy of the parameter estimates depends on the location of persons relative to items (cf. Section 3.6). The person item map is a visual representation of the locations of the items and persons on the latent continuum. In order to be able to estimate the item parameters accurately from the person sample and vice versa, the item difficulties should cover the entire spread of the persons' abilities and vice versa.

An example where this is the case is the left panel of Figure 4.1. For each of the two panels presented in this figure, the relative frequencies of the estimated person parameters are shown at the top, and the locations of the estimated item parameters are shown at the bottom. As we have pointed out previously, both types of parameters are placed on the same latent scale, which is displayed on the x-axis. The item numbers are displayed on the y-axis.

We see that the abilities of the person sample (top part of Figure 4.1, left) are rather symmetrically distributed and stretch over the entire range of the item difficulties (bottom part of Figure 4.1, left). This is an ideal(ized) scenario. For this plot, the person parameters were simulated from a normal distribution. We see that in this example most persons have an average ability, which allows us to estimate the item difficulties in the middle with the highest accuracy, but there are also some persons with very low and very high abilities, which allows us to estimate the more extreme item parameters with some precision as well.

It can also happen, however, that the locations of items and persons are not too well aligned, like in the right panel of Figure 4.1. Here we simulated persons with a higher average ability, but the items' locations are not well suited for measuring these higher abilities on the right side of the continuum.[1] We can also see that there are only few persons at the location of the majority of items towards the left side of the continuum. The locations of persons and items do

[1] Note that in principle the distribution of the estimated person parameters in the right panel of Figure 4.1 should also appear symmetrical, since the true person parameters were sampled from a normal distribution. However, because there are no items on the right side of the continuum, the abilities of those persons on the right cannot be further distinguished.

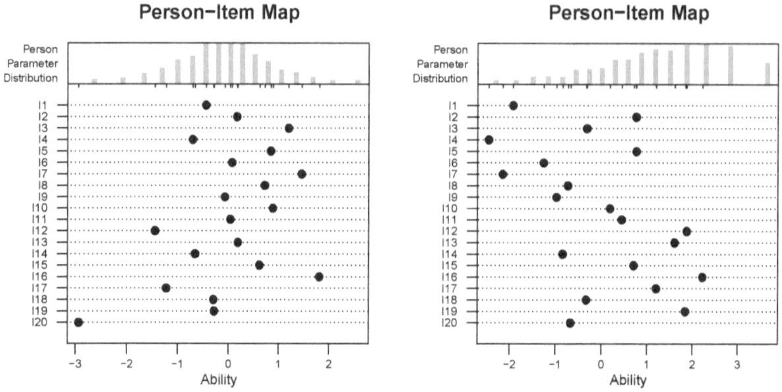

FIGURE 4.1: Person item maps with a good (left) and bad (right) match between the abilities of the test takers and the difficulties of the test items.

not match well in this example, a situation sometimes called mistargeting. If this is the case, the items are not well suited for measuring the abilities of some test takers and at the same time the person sample is not well suited for estimating the item parameters for this test and evaluating its properties.

The person item maps displayed in Figure 4.1 were created by means of the `eRm` package in R, and we will show how this can be done in Section 6.2.1. There and in the following sections we will also see that when we look at the precision of the parameter estimates, both for items in locations with few persons and for persons in locations with few items, the standard errors of the estimates are larger, indicating that the data provide less information for these estimates.

An important distinction pointed out by Andrich and Marais (2019) is that the person sample we would like to score with a test is not necessarily ideal for evaluating the test. An ideal sample for estimating the item parameters and conducting statistical checks on them would contain persons with uniformly spread abilities, so that we would have the same amount of information on item parameters all over this range. Random samples of persons, however, often have unimodal distributions, providing more information on items in the middle of the range.

4.1.2 Empirical ICCs

The ICCs we have presented in Chapter 2 describe the theoretical relationship between the ability of test takers and the probability of a correct response that we expect under the Rasch model for a given difficulty. The expected ICC for an item can be plotted after its difficulty has been estimated. Examples of expected ICCs for four items are depicted in Figure 4.2 as solid curves.

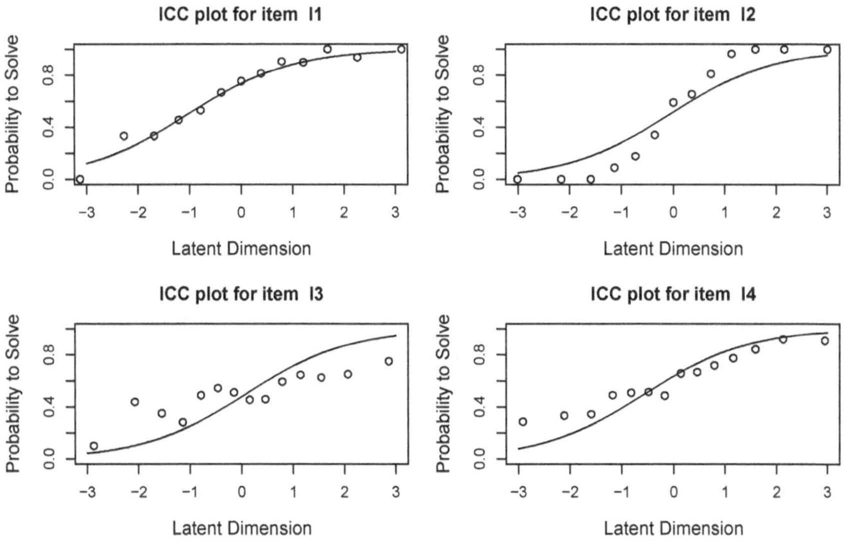

FIGURE 4.2: Expected and empirical ICCs for simulated items showing good fit to the Rasch model (top left), overfit (top right), underfit (bottom left), and guessing (bottom right).

In addition to the expected probabilities of a correct response depicted by the ICC, we can also plot the empirical relative frequencies of a correct response. These empirical relative frequencies are indicated in Figure 4.2 as dots and are called the empirical ICC. We find for the first panel of Figure 4.2 that the empirical relative frequencies fit the theoretically expected shape of the ICC well, whereas for the other panels we see different types of misfit.

In the second panel, the empirical relative frequencies show a steeper slope than expected. This is also termed overfit in the context of the Rasch model.[2] In the third panel, the empirical relative frequencies show a flatter slope than expected. This is also termed underfit in this context. While the Rasch model expects equal slopes, and both under- and overfit can be a reason for concern from the measurement point of view, from a modeling point of view different slopes can be described by the 2PL model presented in Section 4.5.1.1.

In the last panel, the empirical relative frequencies for test takers with low abilities are larger than expected. This pattern is often seen in multiple choice items, where even test takers with a low ability have a chance to pick the correct answer by chance. A model that introduces an extra guessing parameter to account for this pattern is presented in Section 4.5.1.2.

[2]Note that outside of this context the term overfitting refers to statistical or machine learning models that adapt too closely to the training data and thus do not generalize well to new data.

The data used in Figure 4.2 has been simulated for demonstration purposes. Bond and Fox (2007) point out by means of an empirical example that the visual inspection of empirical ICCs should be assisted with fit statistics, which aim at quantifying the acceptable degree of deviation from the Rasch model. A collection of such statistics for quantifying misfit are presented in Section 4.3. The ICCs in Figure 4.2 were created by means of the `eRm` package in R and we will show how to obtain them in the second part of the book (Section 6.2.2).

4.1.3 Graphical Test

The graphical model test (based on Rasch, 1960) is a very intuitive way to evaluate the invariance of test items. It compares the estimated item parameters for two groups of people. For the Rasch model to make sense, the estimates of the item parameters should agree up to a linear transformation. This means that, when graphed, the estimates of the two groups should fall on a straight line.

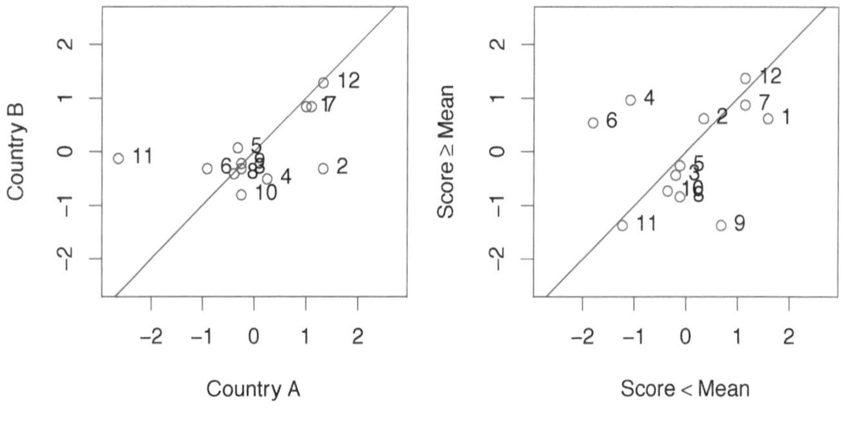

FIGURE 4.3: Comparison of item parameter estimates when item parameters are estimated separately for each country (left) and separately for test takers scoring above and below the average test score (right).

The graphical test is illustrated by Figure 4.3, which shows how item parameter estimates differ when estimated using different subsamples of people. For example, the left panel of Figure 4.3 plots the item parameter estimates for test takers from country A against the item parameter estimates for test takers from country B. In this plot, each point corresponds to the estimates of a single item, whose label appears to the right of the point. The x-coordinate of the point shows its item parameter estimate from the subsample from country A, and the y-coordinate of the point shows its item parameter estimate from the subsample from country B.

We can use Figure 4.3 to assess the items by looking at how far the different pairs of item parameter estimates fall from a straight line. In the Rasch model, an item's difficulty is not supposed to vary across people. Thus, the x- and y-coordinates of each point should be equal (or differ by a constant value), and the point should fall on the reference line $y = x$ (or another straight line, cf. Section 4.2.4).

In practice, item parameter estimates obtained from different samples of people will never be exactly equal, because sample estimates are noisy. This means that small deviations from a straight line are normal, even when test taking behavior follows the Rasch model. Large deviations, however, suggest systematic differences between the two groups. For example, in the left panel of Figure 4.3 Item 11 falls far above the reference line. This suggests that, relative to the other items, this item is easier for men than it is for women. By contrast, Item 2 falls far below the reference line, suggesting that this item is easier for women than it is for men. These are examples of items displaying DIF.

The scatterplots in Figure 4.3 do not indicate how far from the reference line a point needs to be in order to be inconsistent with chance variation under the Rasch model. This can be answered using control lines (Wright & Stone, 1999) or confidence ellipses (Mair et al., 2021). Figure 4.4 adds confidence ellipses to Figure 4.3. The length of each ellipse's horizontal axis shows a 95% confidence interval for the estimate associated with its x-coordinate. In the left panel of Figure 4.4, for example, the horizontal axis shows 95% confidence intervals for the item parameter estimates for test takers from country A. The length of each ellipse's vertical axis shows a 95% confidence interval for the estimate associated with its y-coordinate. In the left panel of Figure 4.4, for example, the vertical axis shows 95% confidence intervals for the item parameter estimates for test takers from country B.

In principle, it is straightforward to use confidence ellipses: When an item's confidence ellipse does not cross the reference line, the two groups have significantly different values for that item parameter, which means that the Rasch model cannot account for test taker behavior. Examples of significant deviations in the left panel of Figure 4.4 are Items 2 and 11. Examples of significant deviations in the right panel of Figure 4.4 are Items 4, 6, and 9. Note, however, that these interpretations depend on the underlying anchoring approach, as will be discussed in Section 4.2.4.

Another inherent problem with all statistical model checks is that even small deviations can be statistically significant for large sample sizes. This problem also affects the confidence ellipses for the graphical test. The larger the sample, the smaller the confidence ellipses. Thus, significant deviations from the reference line will happen increasingly often for larger samples.

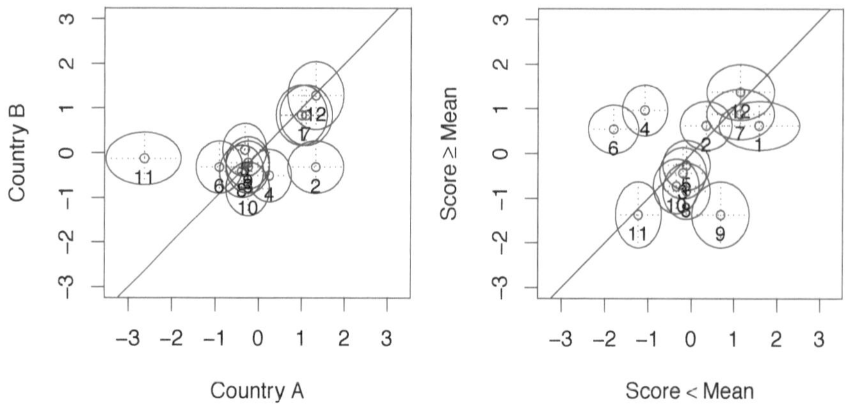

FIGURE 4.4: Comparison of item parameter estimates when item parameters are estimated separately for each country (left) and separately for test takers scoring above the average test score and below (right), including confidence ellipses.

The same principle holds for all of the model checks based on statistical significance that we will cover in the remainder of this chapter. In each case, for a given significance/confidence level the null hypothesis that the Rasch model holds is rejected more readily for larger samples. This also means that, as is the case generally in statistics, a nonsignificant result can also be due to a too small sample size. This is a question of statistical power, which we have already touched upon in Section 3.7.

The graphical tests displayed in Figure 4.3 and Figure 4.4 were created with the `eRm` package in R. We will show how to do this in Section 6.2.3.

4.2 Tests for Item and Person Invariance

We will now present tests for the invariance of item parameters across groups of people and for the invariance of person parameters across groups of items. Under the Rasch model, invariance should hold either way. These tests are based on the following reasoning: Our data consist of several subsets, containing the responses of specific groups of test takers, or the responses to specific groups of items. We now compare two models: Our first candidate model is a single, joint Rasch model, that uses the same item and person parameters for describing the response data in each subset. For our second candidate model, we assume that the Rasch model holds in each of the subsets, but the item or person parameters do not necessarily agree across the subsets. If the first, joint model describes our data as well as the second, more flexible one, we can assume that the item and person parameters are invariant. Otherwise, we detect a violation of this invariance. It is important to note that these tests typically assume that at least the second model is true (i.e., that the Rasch model holds in each subset) and can lead to incorrect results otherwise (Glas & Verhelst, 1995).

4.2.1 Andersen's Likelihood Ratio Test

Andersen's (1973) likelihood ratio test[3] is among the best-known tests for checking whether the Rasch model provides a reasonable account of test-taking behavior.

The likelihood ratio test shares and extends the idea behind the graphical test to more than two groups. When the Rasch model holds, the item parameter estimates for the different groups should not systematically vary. Suppose we split the test takers into groups. This can be done in a variety of ways. For example, Andersen (1973) originally proposed splitting the people into groups by their sum scores, while Gustafsson (1980) split people into sociodemographic groups as a test for DIF. If we now estimate the Rasch model separately for each group, the resulting estimates should be approximately the same, like we discussed already for the graphical test. If they are not, this suggests that the Rasch model does not provide a good account of the test-taking behavior observed.

The likelihood ratio test does not compare the item parameter estimates for the different groups directly. This is done in the graphical test or the item-wise Wald test, which we will cover in Section 4.2.3. Instead, the likelihood ratio test compares the maximum of the conditional likelihood (see Section 3.2) under the Rasch model, where the item parameters must be the same, to the maximum of the conditional likelihood when the item parameters are allowed to vary across groups. These likelihoods provide an index of how well each model accounts for test taker behavior.

When the Rasch model holds, all groups will have the same true item parameters. In this case, the item parameters estimated from all of the data and the item parameters estimated individually for each group will provide similar

[3]Note that there are different kinds of likelihood ratio tests for the Rasch model. In this section, we describe Andersen's (1973) likelihood ratio test, which was the first statistical test suggested for the Rasch model (van der Linden & Hambleton, 1997) and uses all items at the same time. A different likelihood ratio test strategy for DIF detection, which is repeated for individual item parameters consecutively, was suggested by Thissen, Steinberg, and Wainer (1988) and thus abbreviated by De Ayala (2009) as the "TSW" likelihood ratio test. In R it is available in the `difR` package (Magis, 2020). In Section 4.5.2 we will also show how likelihood ratio tests can be used for model comparisons.

accounts of test taker behavior. Consequently, the maximum conditional likelihoods will be approximately equal.

If, on the other hand, the Rasch model is violated and the groups do not have the same true item parameters, allowing each group to have a different item parameter estimate will provide a better account of behavior. In this case, the maximum conditional likelihood obtained by estimating the item parameters separately for each group will be larger than the maximum conditional likelihood obtained by estimating them together. The larger the difference between the true item parameters, the better the account the model with separate item parameters will provide.

The likelihood ratio is the ratio between the conditional likelihood when the item parameters are estimated together and the conditional likelihood when the item parameters are estimated separately. Define $\boldsymbol{u}^{(g)}$, $\boldsymbol{r}^{(g)}$, and $\hat{\boldsymbol{\beta}}^{(g)}$ respectively to be the responses, sum scores, and item parameter estimates for the people in group g, and G to be the overall number of groups. The likelihood ratio is

$$\text{LR} = \frac{L_{\boldsymbol{u}}(\boldsymbol{r}, \hat{\boldsymbol{\beta}})}{\prod_{g=1}^{G} L_{\boldsymbol{u}^{(g)}}(\boldsymbol{r}^{(g)}, \hat{\boldsymbol{\beta}}^{(g)})}.$$

If the item parameters are the same for each person, it will not matter whether we estimate the item parameters together or separately. Except for noise, the estimated values would be identical, and the product of the conditional likelihoods for the groups in the denominator will be identical to the conditional likelihood of the Rasch model in the numerator.[4] This means that the numerator and denominator will be equal. Thus, the null hypothesis for the likelihood ratio test implies that the likelihood ratio will be 1.

By contrast, if one or more groups of people have different item parameters, estimating the item parameters separately for each group will provide a better account of the data. In this case, the product of the conditional likelihoods for each group in the denominator will be greater than the Rasch conditional likelihood in the numerator. Thus, the alternative hypothesis for the likelihood ratio test implies that the likelihood ratio will be less than 1.

The likelihood ratio test does not use the likelihood ratio directly, because it has a complicated distribution under the null hypothesis. Instead, it uses the test statistic $T = -2 \cdot \log(\text{LR})$.[5] For large samples, the sampling distribution of T is approximately $\chi^2(df)$, where the degrees of freedom $df = G \cdot (I-1) - (I-1) = (G-1) \cdot (I-1)$. Under the null hypothesis, where $\text{LR} = 1$, the test statistic $T = 0$ (because $\log(1) = 0$, cf. Appendix A). Under the alternative hypothesis, however, where $\text{LR} < 1$, the test statistic T can take on large values (because the logarithm of small numbers < 1 can give large negative

[4] Because people's responses are independent, the conditional likelihood for any group of people is equal to the product of the conditional likelihoods of the individual people in that group. Since every person is contained in exactly one group, the product of the conditional likelihoods of all groups will be equal to the conditional likelihood for the Rasch model whenever all groups have the same item parameters.

[5] Remember that throughout this book, we use log for the natural logarithm, also denoted as ln.

numbers, but the test statistic is minus two times the logarithm, turning them into large positive numbers).

We can use this to define a statistical test: Values of LR < 1, or large values of T respectively, are unlikely under the null hypothesis. As a consequence, their associated p-values are small, indicating a violation of the Rasch model.

As Glas and Verhelst (1995) point out, for likelihood ratio tests it should also be noted that if the more general model does not describe the data well, a nonsignificant result cannot be considered as support for the restricted model. For Andersen's (1973) likelihood ratio test this means that if the Rasch model does not describe the data well in each group, a nonsignificant test result cannot be seen as support that one joint Rasch model for all groups describes the data well.

We should also be aware that for the likelihood ratio test, the graphical test, and those tests described in the following, the tests' ability to detect group differences is only high when the specified groups are actually the ones that differ in their model parameters. A nonsignificant test result only refers to these particular groups. It does not mean that differences with respect to other grouping variables or specifications could also be ruled out. More flexible approaches for detecting parameter differences are discussed in Section 10.2.

Andersen's likelihood ratio test is available in the `eRm` package in R (see Section 6.2.3).

4.2.2 Martin-Löf Test and Other Approaches for Detecting Multidimensionality

We have seen in the previous section that Andersen's (1973) likelihood ratio test checks the hypothesis that the item parameters are invariant for various groups of persons. A related hypothesis is whether the person parameters are invariant for various groups of items.

Here, the underlying question is whether various groups of items measure different latent traits. This would be a violation of the Rasch model, which implies a single latent trait underlying all items. If this type of model violation is detected, a multidimensional IRT model (see Section 10.3) might be more appropriate (Chalmers, 2021; Reckase, 2009). A common method for assessing dimensionality in general is exploratory factor analysis (with parallel analysis for determining the number of dimensions; see, e.g., Finch & French, 2015; Reckase, 2009).[6] An approach for detecting unaccounted multidimensionality specifically in the Rasch model is to use factor or principal component analysis on the Rasch residuals (Bond & Fox, 2007).

[6] We agree with Reckase (2009) that it is important to point out here that dimensionality is not a property of the test alone, but of a combination of the test itself and the test takers coming from a certain population. For instance, if the test takers do not vary in a certain dimension, it cannot be detected in the items. As a simplified example, imagine that in a sample of patients with major depression a test measuring anxiety may appear unidimensional. However, some of the same test items might turn out to be sensitive to the additional dimension depression in a sample more heterogenous with respect to this dimension.

We will now describe in more detail the Martin-Löf test (Glas & Verhelst, 1995; Gustafsson, 1980), that addresses the alternative hypothesis that groups of items measure different latent traits and is available in the R package eRm. Like Andersen's (1973) likelihood ratio test, this test is based on comparing two conditional likelihoods. The first conditional likelihood $L_{\boldsymbol{u}}(\boldsymbol{r}, \boldsymbol{\beta})$ is that of the Rasch model. The second conditional likelihood $L_{\boldsymbol{u}}(\boldsymbol{r}_1, \boldsymbol{r}_2, \boldsymbol{\beta})$ is again that of a more general model that now allows different person parameters for specific item groups. The item groups need to be defined before the analysis, which can be done based on their difficulties (that is, we test easy against difficult items) or based on different latent dimensions the item groups are suspected to measure (that is, item group 1 is suspected to measure a different latent dimension than item group 2). If the second likelihood is larger, this indicates a violation of the Rasch model (analogously to Andersen's likelihood ratio test).

We now present the test statistic of this test based on the presentation of Rost (2004). We consider a test of I items that belong to two groups with I_1 and I_2 items. Let P be the number of overall respondents, and p_r the number of respondents with a sum score of r. We further use $p_{r_1 r_2}$ to denote the number of respondents that obtained a sum score of r_1 in the first item group and a sum score of r_2 in the second item group. The test statistic of the Martin-Löf test is now given by

$$T = -2\log\left(\frac{\prod_{r=0}^{I}(\frac{p_r}{P})^{p_r}\cdot L_{\boldsymbol{u}}(\boldsymbol{r}, \boldsymbol{\beta})}{\prod_{r_1=0}^{I_1}\prod_{r_2=0}^{I_2}(\frac{p_{r_1 r_2}}{P})^{p_{r_1 r_2}}\cdot L_{\boldsymbol{u}}(\boldsymbol{r}_1, \boldsymbol{r}_2, \boldsymbol{\beta})}\right).$$

Under the Rasch model, this statistic approximately follows a χ^2 distribution with $I_1 \cdot I_2 - 1$ degrees of freedom. Large values of the test statistic, or small p-values, indicate a violation of the Rasch model.

The Martin-Löf test is often described as a test for unidimensionality. As already discussed in Section 2.4.4, certain types of multidimensionality can also show as DIF. For this reason, the other tests described in this section, that aim at detecting DIF, can also be sensitive to certain violations of unidimensionality.

The Martin-Löf test is also available in the eRm package in R (see Section 6.2.7).

4.2.3 Wald Test

We now come back to another test that can be used to check the hypothesis that the item parameters are invariant for various groups of persons, like Andersen's (1973) likelihood ratio test: the Wald test.

The setups of Andersen's (1973) likelihood ratio test and the Wald test are very similar. Both tests are based on the idea that the Rasch model is only a reasonable model for test data if the estimated item parameters do not systematically vary between groups of people. In both tests, we consider estimates of the item parameters for each group of people. Unlike the likelihood ratio test, however, the Wald test compares the groups' item parameter estimates directly. Essentially, the Wald test computes the difference between the first group's estimate of the difficulty of item i, $\hat{\beta}_i^{(1)}$, and the second group's estimate, $\hat{\beta}_i^{(2)}$. This difference is divided by its standard error to account for the fact that all estimates are subject to noise. This results in the test statistic for item i

$$T_i = \frac{\hat{\beta}_i^{(1)} - \hat{\beta}_i^{(2)}}{\sqrt{\text{se}(\hat{\beta}_i^{(1)}) + \text{se}(\hat{\beta}_i^{(2)})}},$$

where $\text{se}(\hat{\beta}_i^{(1)})$ and $\text{se}(\hat{\beta}_i^{(2)})$ denote the standard errors of $\hat{\beta}_i^{(1)}$ and $\hat{\beta}_i^{(2)}$, respectively.

For large samples, T_i will approximately follow a standard normal distribution under the null hypothesis that the true item parameter is the same for both groups. Extreme values of T_i are unlikely under the normal distribution (cf. Appendix B.2). Thus, an extreme value of T_i, with a small p-value, indicates that item i violates the Rasch model.

The Wald test is available in R in `eRm` and `psychotools` (see Section 6.2.4 and Section 6.2.5).

4.2.4 Anchoring

Before moving on, it is important to point out a difficulty that arises when comparing the estimates of individual items. This issue affects both the item-wise Wald test and the previously shown graphical test, as well as any other test comparing individual item parameter estimates. Suppose that we would like to compare each of the item parameters for two groups of people. As noted in Section 2.4.5 and Section 3.2, estimating each group's item parameters requires restricting the item parameters. For example, we might fix the first item parameter to be zero.

When we would like to compare the estimates from two groups, we need to apply a constraint in each group. For example, we could fix the first item parameter to be zero for both groups. In this context, the first item now serves as an anchor item. By setting the first item parameter to the same value in both groups, their item parameters are all placed on the same latent scale.

Anchor items must be selected carefully. If an item parameter is set to the same value in both groups, we can no longer test whether it differs across groups. Figure 4.5 illustrates this problem. In the left panel of Figure 4.5, the first item is used as an anchor. In this case, we can test for group differences for all the other items, and we do find a group difference for the second item. Suppose now that we had used the second item as an anchor instead. This shifts all of the item parameter estimates for the second group, as shown in the middle panel of Figure 4.5. Now when we test for group differences in the

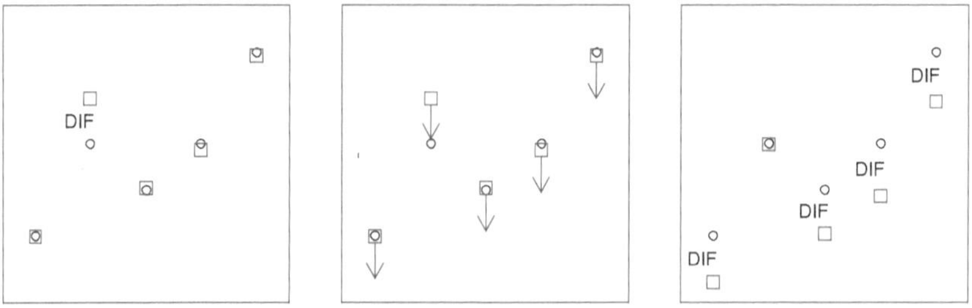

FIGURE 4.5: Illustration of the effect of different anchor item choices on the estimated differences for five item parameters between two groups. The first group's estimates are displayed using circles and the second group's using squares. Differences when the first item is used as the anchor item (left). Demonstration of how using the second item as an anchor instead of the first shifts all item parameter estimates for the second group (middle). Differences when the second item is used as the anchor item (right).

item parameters, we find differences in all but the second item. This is shown in the right panel of Figure 4.5.

Mathematically, the two solutions presented in the left and right panel of Figure 4.5 describe the data equally well. The reason why we typically find the first solution more appropriate is that we find it plausible that a minority of items has DIF, not the majority. This is an assumption shared implicitly or explicitly by many anchoring approaches.

In practice, we do not know a priori which item parameters are invariant across groups. Various approaches for selecting anchor items in a data-driven way have been proposed in the literature (for an overview, see, e.g., Kopf, Zeileis, & Strobl, 2015b). Some of these approaches try to select only invariant items into the anchor, while others try to exclude DIF items from the anchor, which is called purification.

Yet many software implementations of DIF tests by default use the constraint that the sum of all item parameters be zero in both groups. This is the case, e.g., in the `eRm`[7] and `difR` packages in R, but both packages also offer means for stepwise item elimination or purification. The sum zero constraint corresponds to an anchor containing all items and implicitly assumes that DIF is balanced and will cancel out across all items. If this is not the case, however, or generally in situations where the anchor contains DIF items, other items can artificially appear to exhibit DIF (as we have seen in Figure 4.5; see also, e.g., Fischer & Molenaar, 1995; Wang, Shih, & Sun, 2012).

[7]For example, the graphical test from the `eRm` package, that we have seen in Section 4.1.3, uses this constraint. In the exercises to this section we will show an example further illustrating this point.

In Section 6.2.5 we will illustrate the practical application of the stepwise approach available in the `eRm` package as well as additional modern anchoring approaches available in the R package `psychotools` that do not rely on the assumption that DIF is balanced.

4.2.5 Other Approaches for Detecting DIF

In Part III of this book we will discuss extensions of the Rasch model that allow different groups of people to have different item parameters. The mixture Rasch model (Rost, 1990) is the best-known model of this type. This model can be used to look for unobserved groups of test takers, each of which having their own set of item parameters. This is similar to the setup for the likelihood ratio and Wald tests. In the likelihood ratio and Wald tests, however, we have specific hypotheses about which groups may exhibit DIF. In the mixture Rasch model, test taker groups are learned from the data. Consequently, we can use the mixture Rasch model and another approach described in Section 10.2 to check for DIF when we do not have concrete hypotheses about which groups of test takers exhibit DIF.

We would also like to mention that several other statistical tests have been proposed to detect DIF without employing the Rasch model (or more general models from IRT). The most prominent examples of these "model-free" DIF tests are logistic regression and the Mantel-Haenszel test (cf., e.g., Holland, Thayer, Wainer, & Braun, 1988; Magis, Béland, Tuerlinckx, & De Boeck, 2010). In this book we focus on DIF detection approaches within the Rasch model framework. However, these methods are very well explained by Finch and French (2015), who illustrate their practical application with the `difR` package and also emphasize that the scores used for matching need to be purified in order to avoid contamination with DIF items.

4.2.6 How to Proceed with Problematic Items

The aim of the statistical approaches we have just described is to find out whether certain items function differently between certain groups of test takers. Once DIF items are detected, there are different ways to proceed, and which one will be taken depends on different aspects, including the cost of creating the items in the first place. Often DIF items are simply excluded from the test. Sometimes, when the DIF is detected at an early stage of test development, it may also be possible to modify an item. For example, if a DIF item contains a complicated or rarely used term and turns out to be harder to answer for non-native speakers of the test language, it might be worth a try to adapt the phrasing. Then the test would have to be administered and evaluated again to see whether the attempt to remove the suspected source of DIF has been successful.

With respect to the notion of fairness, Camilli (2006) emphasizes that statistical evidence of DIF provides no proof that an item is unfair. Only if an item shows DIF and the source of this DIF is irrelevant to the construct the test is intended to measure, should it be termed unfair. This is why we argued in the introduction to this book that while it may be legitimate that reading skills affect the results of a test on college readiness, they should not affect the results of a test that claims to measure only algebra proficiency. In this latter example, items disadvantaging non-native speakers need to be modified or excluded to establish a fair test.

Another possible approach to deal with DIF is to "split" the item in the further analysis of the existing test. For this, two "copies" of the item are used to encode the data. The main idea is now to treat the item as two different items in the two groups, which are therefore allowed to differ with regard to their psychometric characteristics. Of the two groups between which the item shows DIF, one group receives their responses to the item in the column for the first copy and missing values in the column for the second copy of the item. The second group receives their responses to the item in the column for the second copy and missing values in the column for the first copy of the item. This acts as if the two groups had seen two different items, whose item parameter estimates can differ between the groups. This general approach is also common in factor analysis and referred to as partial measurement invariance.

Similarly, how to proceed with items that show problems in any of the other evaluation approaches depends on the stage of test development, considerations about item content, and similar aspects. As Andrich and Marais (2019, p. 193) point out with respect to item fit indices, "[...] multiple pieces of evidence should be used in making any decision to modify, discard or deal with an item in any way."

4.3 Goodness-of-Fit Tests and Statistics

In this section we review a variety of approaches for evaluating the fit between item response data and the Rasch model. Some of them are formal statistical tests while others are descriptive statistics for which rule-of-thumb critical boundaries have been suggested in the literature. We will also see that there are both approaches that evaluate model fit at the level of the entire psychological test as well as approaches that evaluate the fit of single items or persons.

4.3.1 χ^2 and G^2 Goodness-of-Fit Tests

We will now introduce a second class of goodness-of-fit tests, which are also widely used in the analysis of contingency tables (Agresti, 2002) and were described in the context of item response theory, for instance, by Maydeu-Olivares and Montaño (2013). The χ^2 goodness-of-fit test takes a different approach than, for example, the likelihood ratio or Martin-Löf tests presented in the previous section, in that it is not based on comparing the relative fit of two models. Instead, a χ^2 test checks how well the response patterns predicted by the Rasch model match the response patterns we observe from our test takers. It does this by comparing the observed number of test takers showing each response pattern to the number that we would expect under the Rasch model (cf. Appendix B.2.1 for an introduction to the general principle of χ^2 tests).

We start by presenting the general idea of χ^2 goodness-of-fit tests for the Rasch model, which will be refined later. For this, we consider all possible response patterns (i.e., all possible sequences of 0s and 1s for incorrect and correct responses) that can result from answering the test items. Let $O_{\boldsymbol{u}}$ be the observed number of test takers responding with pattern $\boldsymbol{u}$ and $E_{\boldsymbol{u}}$ be the number of test takers we would expect to respond with pattern $\boldsymbol{u}$ under the Rasch model. Then, the test statistic for the χ^2 test is

$$T = \sum_{\boldsymbol{u}} \frac{(O_{\boldsymbol{u}} - E_{\boldsymbol{u}})^2}{E_{\boldsymbol{u}}}.$$

Note that in this expression, the differences are squared, and as a consequence the positive and negative differences in the sum will not cancel out. Then, each difference is weighted by the inverse of the expected frequency, since large frequencies will tend to also result in larger differences. In very large samples, the test statistic T will approximately follow a χ^2 distribution when the Rasch model (or any other model from which the $E_{\boldsymbol{u}}$ values are derived) holds. Thus, observing a large value of T, or equivalently a small p-value, suggests a poor fit.

A problem of this approach is that the χ^2 goodness-of-fit test is only valid if the expected frequency of each possible response pattern is sufficiently large, which is a general requirement for χ^2 tests. However, for tests with many items, the number of possible response patterns will become so large that we expect some of the patterns to rarely occur. In this case, the test statistic T will not follow a χ^2 distribution under the null hypothesis. For this reason, the χ^2 goodness-of-fit test in the form we have shown above for didactic reasons is typically not applied in practice.

Instead of using individual response patterns, it is possible to combine response patterns to obtain higher expected frequencies for groups of responses, e.g. by combining responses for predefined ability groups, as we will demonstrate below. This solves the problem with the rare patterns and allows the definition of goodness-of-fit statistics for both items and persons. An alternative is limited information approaches, which are presented in the next section.

Another classical goodness-of-fit statistic, which also stems from the analysis of categorical data, is the likelihood ratio statistic G^2. Using the notation from above, this statistic is calculated as follows (Maydeu-Olivares, 2013):

$$G^2 = \sum_{\boldsymbol{u}} 2O_{\boldsymbol{u}} \log \left(\frac{O_{\boldsymbol{u}}}{E_{\boldsymbol{u}}} \right).$$

As its name suggests, the reasoning behind G^2 is similar to that of a likelihood ratio test, but it compares observed and expected frequencies instead of the likelihoods of two models. If the expected frequencies for the response patterns are close to the observed frequencies, the ratio $\frac{O_{\boldsymbol{u}}}{E_{\boldsymbol{u}}}$ is close to 1 for every response pattern, and the natural logarithm $\log(\frac{O_{\boldsymbol{u}}}{E_{\boldsymbol{u}}})$ is close to 0, leading to a G^2 statistic close to 0 if the model fits well. Like T, G^2 approximately follows a χ^2 distribution when the model leading to the $E_{\boldsymbol{u}}$ values (for instance, the Rasch model) holds.

Like for the χ^2 goodness-of-fit test, the p-values for G^2 are only valid for large expected frequencies (Maydeu-Olivares, 2013). As a result, G^2 is also typically not used in practice, but was presented here because other test statistics introduced below are based on it.

4.3.2 M_2, RMSEA, and SRMSR

We can get around the problem of rare response patterns by employing Maydeu-Olivares and Joe's (2006) M_2 statistic. Rather than comparing the frequencies of entire response patterns, the M_2 statistic uses the information contained in the responses to individual items and to item pairs. Thus, the M_2 statistic compares a) the expected and observed frequencies for correct responses to individual items, and b) the expected and observed frequencies for correct responses to both items of an item pair. For instance, for the first and second item, it compares the observed and expected frequencies for a correct response to the first item, to the second item, and to both items. Conceptually, this procedure can be compared to investigating frequency tables for item pairs (Maydeu-Olivares, 2013).

While this results in considering a large number of individual items and item pairs, the expected frequencies for correct responses to single items or to both items in an item pair will generally be large enough to make them more amenable to the idea of χ^2 testing. Like for the χ^2 goodness-of-fit test, a large value of M_2, or equivalently a small p-value, suggests a poor fit between the data and the Rasch model. It can further be shown that, if no model violation is present, the M_2 statistic approximately follows a χ^2 distribution with $k-d$ degrees of freedom. Here, k is the number of the frequencies compared by the statistic, that is the number of items and item pairs, and d is the number of free model parameters (Maydeu-Olivares, Cai, & Hernández, 2011).

The M_2 statistic can also be used to calculate another overall goodness-of-fit statistic, the root mean square error of approximation (RMSEA). Using the M_2 statistic, its degrees of freedom df, which are again given by $k-d$, and the sample size P, the RMSEA value is calculated as (Maydeu-Olivares et al., 2011)

$$RMSEA = \sqrt{\frac{M_2 - df}{P \cdot df}}.$$

RMSEA values close to 0 typically indicate a good model fit. There seem to be no generally established guidelines on the interpretation of this statistic, but 0.05 has been suggested as a rough threshold for a good model fit (Maydeu-Olivares et al., 2011). The application of this statistic will be demonstrated in Section 7.3.

There are further overall fit statistics available, such as the standardized root mean square residual (SRMSR, Maydeu-Olivares, 2013). The idea of SRMSR and related statistics is to compare the observed correlations or covariances between all item pairs with those predicted under the Rasch model (or another IRT model). Values close to 0 indicate a good model fit and Maydeu-Olivares (2013) recommends a cutoff value of 0.05 for SRMSR.

While the overall fit statistics reported in this section are computed for the entire test, the following sections will present fit statistics for individual items (and persons).

4.3.3 Infit and Outfit Statistics

The underlying idea of the χ^2 and M_2 tests discussed in the previous section is the comparison of observed frequencies with those expected under the Rasch model. The fit statistics presented next are based on a similar approach using Rasch residuals. These are the differences between the observed responses (i.e., the 0 or 1 responses for dichotomous items) and their expected values (i.e., the predicted probabilities for a correct response under the Rasch model) (Wright & Masters, 1990). In the context of the Rasch model, these expected values are typically calculated based on the item and person parameter estimates.

Generally speaking, when there is a good fit between data and model, it can be expected that residuals are small. It thus seems natural that Rasch residuals can be used for assessing the fit of the Rasch model. Below we will see that in Rasch analyses not only the case where the residuals are larger than expected can be a reason for concern, but also the case where the residuals are smaller than expected.

A common approach for checking the fit of individual items using Rasch residuals consists of calculating the infit and outfit statistics. Details on their computation are provided, for instance, by Wright and Masters (1982). We will outline the steps for calculating these statistics before addressing their interpretation. We focus on the case that these statistics are calculated for individual items. By switching the roles of respondents and items, it is also possible to use infit and outfit statistics to check the fit of individual persons.

The calculation of the outfit statistic for a specific item is based on the following steps: First, the Rasch residuals for the responses of each test taker

Outfit and infit statistics		Interpretation
MSQ	t	
$>> 1$	$>> 0$	Underfit
(responses more random than expected, residuals larger than expected)		(empirical ICC too flat)
$<< 1$	$<< 0$	Overfit
(responses less random than expected, residuals smaller than expected)		(empirical ICC too steep)

TABLE 4.1: An overview of the meaning of different values for outfit and infit mean squared (MSQ) and standardized (t) statistics, where the respective infit statistics are less sensitive to outliers.

to this item are calculated. Second, the Rasch residuals are standardized so that they have a mean of zero and variance of one. Third, the standardized Rasch residuals are squared and summed over all respondents. Fourth, this sum is divided by the number of respondents to obtain the outfit mean square statistic. We can summarize this procedure by the following formula (Wright & Masters, 1990). If Z_{pi} denotes the standardized Rasch residual in the response of respondent p to item i, and P the number of respondents, the outfit mean squared (MSQ) statistic for item i can be calculated as:

$$\text{Outfit } MSQ_i = \frac{\sum_{p=1}^{P} Z_{pi}^2}{P}.$$

Following Wright and Masters (1990), this statistic has an expected value of 1 under the Rasch model. Values above 1 indicate Rasch residuals that are larger than expected under the Rasch model, and thus a possible model violation. Such items are also said to show underfit (Bond & Fox, 2007). Values below 1 indicate that the residuals are lower than expected. This is considered to indicate an overfit of the responses to the Rasch model. In this context, overfit means that the deviation between the expected values and the empirical data is smaller than expected. The interpretation of different values for infit and outfit statistics is summarized in Table 4.1.

We can further obtain a standardized weighted mean square statistic for each item, which is typically denoted by t_i. Let $\sqrt[3]{MSQ_i}$ and $sd(MSQ_i)$ denote the cube root and the expected standard deviation of Outfit MSQ_i, respectively. Then the standardized statistic t_i is

$$\text{Outfit } t_i = (\sqrt[3]{MSQ_i} - 1)(3/sd(MSQ_i)) + (sd(MSQ_i)/3).$$

This standardized statistic t_i is often presented in software output in addition to the MSQ statistic (for the meaning of different values for t_i see again Table 4.1).

Items showing underfit and overfit can also be roughly identified using their empirical ICCs, as we have already seen in Section 4.1.2. Items showing underfit have empirical ICCs that are flatter than those expected under the Rasch model. Items showing overfit have empirical ICCs that are steeper than expected. This connection is further illustrated in Figure 4.6: The top row shows the expected ICCs and item responses, together with the resulting empirical ICCs, for three simulated items. The bottom row shows the same expected ICCs and item responses, together with their Rasch residuals.

We see that for an item with underfit (Figure 4.6, left column) correct and incorrect responses (crosses) are randomly scattered across a wide range of the latent continuum. This results in a flatter empirical ICC (circles, top plot) and in many large residuals (dotted lines, bottom plot), which lead to high values for MSQ and t. For an item with overfit (Figure 4.6, right) correct and incorrect responses (crosses) are almost deterministically separated. This results in a steeper empirical ICC (circles, top plot) and in many small residuals (dotted lines, bottom plot), which lead to low values for MSQ and t. When the Rasch model holds (Figure 4.6, middle) the responses show the degree of randomness we expect from the model: Correct and incorrect responses are equally likely close to the item's difficulty in the middle of the latent continuum, while correct responses are much more likely for high abilities and incorrect responses are much more likely for low abilities. In this case, the empirical ICC (circles, top plot) follows the shape of the expected ICC (solid line) and the residuals match what is to be expected under the Rasch model, which leads to an MSQ value close to 1 and a t value close to 0.

The infit MSQ statistic follows a similar reasoning as the outfit statistic. In contrast to the outfit statistic, however, the Rasch residuals of the individual responses are now weighted with their expected variance under the Rasch model (Wright & Masters, 1990). Therefore, residuals with small variances, which occur for large distances between person and item locations, have comparatively less effect on this statistic. As a consequence, the infit statistic is less strongly influenced by outliers than the outfit statistic (Bond & Fox, 2007; Linacre, 2002). We will see examples where the infit and outfit statistics diverge in the second part of the book.

We again present the respective formula for the infit MSQ statistic. Let Z_{pi} again denote the standardized Rasch residual in the response of respondent p to item i, and P the number of respondents, and let W_{pi} denote the expected variance of Z_{pi} under the Rasch model. The infit MSQ statistic for item i is then denoted by:

$$\text{Infit } MSQ_i = \frac{\sum_{p=1}^{P} W_{pi} Z_{pi}^2}{\sum_{p=1}^{P} W_{pi}}.$$

The equation for t_i can again be used to obtain a standardized version of this statistic for each item, which is correspondingly called the Infit t statistic.

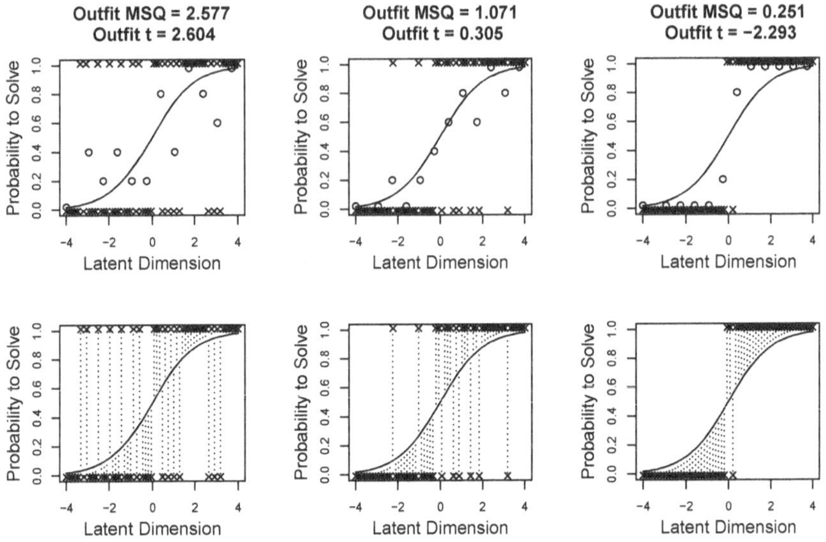

FIGURE 4.6: Top row with responses (crosses at top and bottom), expected ICC (solid line) and empirical ICC (circles); bottom row with responses (crosses at top and bottom), expected ICC (solid line) and Rasch residuals (dotted lines) for simulated items with underfit (left), responses sampled from the Rasch model (middle), and overfit (right).

For both the MSQ and t values of the infit and outfit statistics, various cutoff values have been proposed, and we will list some examples. Bond and Fox (2007) and Engelhard (2013) mention cutoff values of -2 and 2 for the t statistics, whereas Paek and Cole (2020) suggest -3 and 3. Similarly, Bond and Fox (2007) give 0.75 and 1.3 as cutoff values for the MSQ statistics, whereas DeMars (2010) mentions 0.6 and 1.5 as possible alternatives. Desjardins and Bulut (2018), on the other hand, argue against the use of specific cutoff values for these statistics.

In R, infit and outfit statistics are available in the `eRm`, `mirt`, and `TAM` packages. We address their practical calculation in `eRm` and `mirt` in the corresponding chapters in the second part of the book.

The `eRm` package further presents a χ^2-type fit statistic that is closely related to the infit and outfit statistics. This statistic can be calculated for individual items or persons. When it is calculated for individual items, it is the sum of the squared standardized Rasch residuals Z_{pi}^2 over all test takers (i.e., it is equal to the numerator of the outfit MSQ statistic). Analogously, when it is calculated for test takers, it is the sum of the squared standardized Rasch residuals Z_{pi}^2 over all items. If all standardized Rasch residuals would follow a standard normal distribution, the sum of the squared standardized Rasch residuals would follow a χ^2 distribution, with the degrees of freedom equal to number of terms in the sum (i.e., the number of test takers or items, excluding those with a perfect score or a score of zero) minus one. In general, this assumption does not hold in a strict sense, but it is sometimes used as a rough approximation. The statistic becomes large, and the p-value resulting from the χ^2 approximation becomes small, when the sum of squared residuals is large. Due to the asymmetry of the χ^2 distribution (cf. Appendix B.2.1), this means that the χ^2 fit statistic will show significant p-values only for underfitting, not for overfitting items or persons.

4.3.4 Further Fit Statistics for Items

Besides the infit and outfit statistics considered in the previous section, numerous authors have proposed additional fit statistics based on the differences between the observed responses and the values expected under the model. They were proposed not only for individual items, but also for specific item pairs. In this section, we will mention some important examples. These statistics are available in R, and we will discuss their application in the second part of the book.

Orlando and Thissen (2000, 2003) proposed and evaluated several fit statistics for individual items. An important example is $S\text{-}X_i^2$. It is calculated as follows: First, the observed sum scores are divided into $I-1$ ability groups, where I is the number of items. These ability groups correspond to respondents with the same number of correct responses, with the least able group having one correct response, and the most able group having $I-1$ correct responses. For each ability group answering r items correctly, the number of respondents in this group, P_r, the proportion of observed correct responses to item i in this ability group, O_{ir}, and the expected proportion based on the Rasch model, E_{ir}, are calculated. Now $S\text{-}X_i^2$ can be calculated by the formula:

$$S\text{-}X_i^2 = \sum_{r=1}^{I-1} P_r \frac{(O_{ir} - E_{ir})^2}{E_{ir}(1 - E_{ir})}$$

Small values indicate a good fit, in that the expected value is close to the observed value for each ability group. Although the formula for this statistic resembles that for the χ^2 test, its distribution is not always well described by a χ^2 distribution, particularly in short tests. As a consequence, this statistic can have a slight tendency to detect item misfit in small tests even when the IRT model fits well (Orlando & Thissen, 2000). This statistic is sometimes also called the signed χ^2 statistic (Desjardins & Bulut, 2018).

A closely related test statistic is Q_1 (Yen, 1981), which was also named $Q_1 - X^2$ by Orlando and Thissen (2000). It uses the same terms as $S\text{-}X_i^2$, but is based on a fixed set of 10 ability groups instead of $I-1$. Chalmers and Ng

(2017) proposed an improved version of Q_1, which allows the calculation of p-values for individual items. The calculation of this improved version in `mirt` also takes the uncertainty of the person parameter estimates into account, as we will show in Section 7.3.

Another related test statistic is Bock's (1972) χ^2 method. In its implementation in the `mirt` package, which will be demonstrated in Section 7.3, it is based on the following statistic:

$$\chi_i^2 = \sum_v P_v \frac{(O_{vi} - E_{vi})^2}{E_{vi}(1 - E_{vi})}.$$

This statistic is based on intervals of person parameter values, which have been defined before the analysis. For each interval v, P_v denotes the number of test takers whose person parameter estimate falls into this interval. The expected proportions E_j are estimated based on the ability value in the center of each interval. The idea of this statistic is thus very similar to that of $S\text{-}X_i^2$ and Q_1, which only differ in how the ability groups are defined. Again, a value close to 0 indicates that the expected values are close to the observed values for each ability group.

The idea of using ability groups to obtain larger expected values has also been applied to the G^2 statistic from Section 4.3.1. We present here the G_i^2 statistic, which was described by McKinley and Mills (1985) and is implemented in the `mirt` package. Using the same terms as Bock's χ^2 method, this statistic is calculated as follows:

$$G_i^2 = 2 \sum_v P_v \left(O_{vi} \log\left(\frac{O_{vi}}{E_{vi}}\right) + (1 - O_{vi}) \log\left(\frac{1 - O_{vi}}{1 - E_{vi}}\right)\right).$$

Again, a small value indicates a good agreement of observed and expected values. Although these statistics are designed to follow χ^2 distributions, this is typically not the case in empirical data even when all items show a good fit to the model. However, values close to 0 generally indicate a good fit between the respective item and the model, and values much larger than 0 typically indicate a model misfit. Values below 0 are impossible. To point out its relationship to Q_1, Orlando and Thissen (2000) named a variant of this statistic, which uses 10 ability groups, $Q_1 - G^2$. These statistics are available in R, and we will show their calculation in the `mirt` package in Section 7.3. This software uses the χ^2 distribution to calculate approximate p-values, but again this approximation does not work well for short tests. For instance, Orlando and Thissen (2000) found that G^2, when it uses 10 ability groups, and Q_1 both show a strong tendency to give false-positive results, that is, to report an item that fits well as showing misfit, particularly in short tests.

A more sophisticated approach for testing the significance of model violations for individual items using χ^2 statistics was proposed by Stone (2000). In contrast to the methods presented so far, this method is based on a Bayesian framework and aims at considering the imprecision of the ability estimate on checking the goodness-of-fit of the observed responses. It is included in the `mirt` package and its computation will be demonstrated in Section 7.3.

4.3.5 Fit Statistics for Item Pairs

So far, model residuals were used to evaluate the fit of individual items. Another set of test statistics aims at investigating the relationship between the residuals of item pairs. The underlying idea is to detect relationships between two items that are not expected under the Rasch model. These statistics usually aim at the detection of a violation of the assumption of local independence, which can be related to learning effects, undetected multidimensionality or similar effects. A well-known goodness-of-fit statistic for item pairs is Q_3 (Yen, 1984), which is simply the correlation between the Rasch (or other IRT model) residuals of two items i and j, calculated over all test takers. A high correlation could result, for example, when test takers can learn something from solving item i that helps them solve item j. The distribution of Q_3 is typically unknown even if the model-data fit is good, although a value (that is, a correlation) close to 0 typically indicates that two items are locally independent.

Various cutoff values have been suggested for the Q_3 statistic, which are typically compared to the absolute value of Q_3. According to Christensen, Makransky, and Horton (2017), 0.2 and 0.3 are often used, but these cutoff values might not be suited for all applications. Christensen et al. (2017) also mention an alternative method which compares the value of Q_3 for a specific item pair against its average value observed for all item pairs. In a simulation study, they found that values of Q_3 that lie more than 0.2 above the average value could indicate local dependency. Several methods have also been proposed to obtain critical values for Q_3 or to calculate p-values, for instance by Christensen et al. (2017). Debelak and Koller (2020) proposed a related nonparametric model test for the Rasch model that is based on Q_3-based statistics. This nonparametric test is available in `eRm`, together with further tests that are based on other fit statistics. We provide an overview over the theory and application of nonparametric fit tests in Section 4.3.7 and Section 6.5. However, these tests are only recommended for small data sets, since the computation of the p-values is computationally intensive.

Q_3 as a descriptive statistic is available in the `mirt` and `TAM` packages, as illustrated in Section 7.3 and Section 8.2. In contrast to the nonparametric model tests, the descriptive statistic can also be calculated in larger data sets.

Other test statistics for checking the local independence assumption are X^2 and G^2, which were both proposed by Chen and Thissen (1997). For these statistics, p-values can be computed for testing the null hypothesis that a specific item pair is locally independent. Small p-values thus indicate significant local dependence, i.e., that the responses to two items are more similar or more dissimilar than predicted by the model. X^2 is available in the `TAM` package (cf. Section 8.2).

4.3.6 Fit Statistics for Persons

After discussing the fit of items and item pairs, we are now interested in checking the goodness-of-fit of persons. The aim of the analysis is now to assess response patterns of individual persons that deviate from those expected under the Rasch model. A straightforward approach that is based on Rasch residuals is to calculate infit and outfit statistics for persons. The formulas for this approach correspond to those presented in Section 4.3.3, but the roles of items and test takers are switched. Examples for the interpretation of infit and outfit statistics are given, for instance, by Bond and Fox (2007). Following these authors, overfit of persons can be an indicator for a deterministic response behavior, whereas underfit could indicate sloppiness or lucky guessing. According to Engelhard (2013), values of the infit and outfit t statistics that lie below -2 or above 2 indicate unexpected response behavior. This corresponds to the interpretation guidelines for the same fit statistics for items and is illustrated in Section 6.2.8.

Another important class of person fit statistics is based on the idea of considering test statistics that roughly follow a standard normal distribution if a person shows a good model fit. Values close to 0 indicate a good model fit. Positive values (much) larger than 0 and negative values (much) smaller than 0 both indicate a misfit, but typically they indicate a different type of misfit. An important example is the z_h (or Z_h) statistic (Drasgow, Levine, & Williams, 1985). We will explain its rationale for dichotomous items. As usual, let I be the number of items in our test and let p be an arbitrary person for which we would like to assess the fit to a given model. We now treat the item parameters as known, and obtain ability parameter estimates for person p, which we denote by $\hat{\theta}_p$. We start by defining the following term, where $\Pr(U_{pi} = 1|\hat{\theta}_p)$ is the probability for giving a correct response to item i if person p has the ability parameter $\hat{\theta}_p$ and u_{pi} is the actual response of person p to item i:

$$l_{0p} = \sum_{i=1}^{I} u_{pi} \log \Pr(U_{pi} = 1|\hat{\theta}_p) + (1 - u_{pi}) \log(1 - \Pr(U_{pi} = 1|\hat{\theta}_p)).$$

Given a person parameter estimate $\hat{\theta}_p$ for person p, l_{0p} is the natural logarithm of the log likelihood function for the response pattern of person p (Drasgow et al., 1985). Using the expected value of l_{0p} based on the assumed model and $\hat{\theta}_p$, which we denote by $E(l_{0p})$, and its standard deviation, $sd(l_{0p})$, we obtain

$$z_{hp} = \sum_{i=1}^{I} \frac{l_{0p} - E(l_{0p})}{sd(l_{0p})}.$$

Large negative values, for instance, values below -2, indicate that the responses are overall implausible under the model and therefore indicate a person misfit. Values close to 0 indicate that the response patterns correspond to those expected under the Rasch model. Large positive values indicate response patterns that have a higher likelihood than is predicted under the Rasch (or other IRT) model (Embretson & Reise, 2000, p. 240); such patterns could occur, for instance, when a test taker behaves more deterministically than expected and gives (almost) only correct responses to easy items and (almost) only incorrect responses to difficult items. A similar logic can be applied to obtain a person fit statistic z_h for polytomous items, as well as an item fit statistic for individual items. In `mirt`, z_h is available as an item fit statistic as well as a person fit statistic. We will discuss their calculation in Section 7.3.

4.3.7 Nonparametric Goodness-of-Fit Tests

Finally, a set of nonparametric goodness-of-fit tests was proposed by Ponocny (2001). These tests are based on bootstrapping and are particularly useful to check the model fit in small data sets. So far, these tests are only available for the dichotomous Rasch model when no missing data are present. The reasoning behind these tests can be summarized as follows: We consider a response matrix of 0s and 1s and assume that the Rasch model describes these data well. If this is the case, it can be shown that all response matrices with equal sums in the rows and columns are equally likely (or plausible) under the Rasch model. We now summarize our observed response matrix by a test statistic of interest. If the Rasch model holds, we can expect that our response matrix is not unusual under this model. As a consequence, the value of our statistic can be expected to be close to the values observed in other matrices with the same row and column sums. Otherwise, if our statistic is indeed unusual compared to those obtained for other response matrices with the same row and column sums, this might indicate a model violation. The steps of these tests can now be summarized as follows:

- Select a test statistic, which corresponds to a model violation of interest. This could be, for instance, Q_3 for a specific item pair, or another item-wise or overall fit statistic. Calculate it for the observed response matrix.
- Using a bootstrap algorithm, generate many matrices with the same row and column sums as the original response matrix. Calculate the statistic for these artificial matrices as well. This leads to a distribution of statistics that serves as a reference distribution.
- Compare the observed value with the reference distribution by determining its rank among the ordered values. The percentage of values in the reference distribution that are smaller than the observed value is interpreted as a p-value. Extreme values close to 0 or 1 indicate a model violation.

A number of statistics were proposed for this approach, for instance, T_{10} as a nonparametric counterpart to Andersen's likelihood ratio test (Ponocny, 2001), T_{11} as an overall goodness-of-fit test (Ponocny, 2001) as well as T_1, T_{1l}, T_{1m}, Q_{3h}, and Q_{3l} as local independence tests for item pairs

(Debelak & Koller, 2020; Koller & Hatzinger, 2013; Ponocny, 2001). The application of some of these tests in R will be illustrated in Section 6.5.

A study of Debelak (2019) compared the power of the nonparametric tests based on T_{10} and T_{11} with that of Andersen's likelihood ratio test and M_2 against several model violations of the Rasch model. T_{10} and Andersen's likelihood ratio test used the median of the sum score as the splitting criterion. It was found that the T_{11} test is most powerful against violations of the assumption of parallel item characteristic curves and therefore against more general models, such as the 2PL and 3PL models (cf. Section 4.5.1.1 and Section 4.5.1.2). Against local dependence, M_2 was found to be most powerful. The remaining two tests were found to be sensitive against violations of the assumption of parallel item characteristic curves, but to be insensitive against local dependence.

4.3.8 Posterior Predictive Checks

In Bayesian settings, posterior predictive checks are used to assess goodness-of-fit. Similar to χ^2 tests, posterior predictive checks compare the data predicted by the model to the data observed. However, unlike χ^2 tests, posterior predictive checks incorporate all of the uncertainty contained in the posterior into their predictions. Thus, the more diffuse the joint posterior distribution over $\boldsymbol{\theta}$ and $\boldsymbol{\beta}$, the more diffuse our predictions about the data matrices we expect to observe.

The posterior predictive distribution is $\Pr(\tilde{\boldsymbol{U}} \mid \boldsymbol{U} = \boldsymbol{u})$, the probability distribution over new data matrices $\tilde{\boldsymbol{U}}$ given the data matrix $\boldsymbol{u}$ that we have already observed. In subsequent equations, we will write $\boldsymbol{U}$ instead of $\boldsymbol{U} = \boldsymbol{u}$. We can compute the posterior predictive distribution by noting that, given $\boldsymbol{\theta}$ and $\boldsymbol{\beta}$, the probability of any data matrix is given by the Rasch model. This means that

$$\Pr(\tilde{\boldsymbol{U}} \mid \boldsymbol{\theta}, \boldsymbol{\beta}) = \prod_{p=1}^{P} \prod_{i=1}^{I} \frac{e^{\theta_p - \beta_i}}{1 + e^{\theta_p - \beta_i}}.$$

The posterior predictive distribution is the average of $\Pr(\tilde{\boldsymbol{U}} \mid \boldsymbol{\theta}, \boldsymbol{\beta})$ over the posterior distribution

$$\int_{\boldsymbol{\theta}} \int_{\boldsymbol{\beta}} \Pr(\tilde{\boldsymbol{U}} \mid \boldsymbol{\theta}, \boldsymbol{\beta}) \cdot f(\boldsymbol{\theta}, \boldsymbol{\beta} \mid \boldsymbol{U}) d\boldsymbol{\beta} d\boldsymbol{\theta}.$$

Typically, we do not compute the posterior predictive distribution by hand. Instead, we sample the posterior predictive distribution by sampling a new value of $\tilde{\boldsymbol{U}}$ for every sample from the posterior.

Often we are interested in the posterior predictive distribution of a summary statistic computed from the data matrix, rather than the data matrix itself. Sinharay, Johnson, and Stern (2006) compared different summary statistics on their ability to detect violations of the assumptions of the Rasch model.

In Section 9.3, we will use one of the statistics, the point biserial correlation, to illustrate how the posterior predictive can be used to assess goodness-of-fit.

4.4 Separation Indices

In this section, we present two indices that have been proposed in Rasch measurement to indicate how well items of a psychological test separate between respondents of a population and, vice versa, how well test items are separated by respondents. This question is related to the problem of assessing the reliability of psychological tests and to the estimation of the item and test information, which was addressed in Section 3.6.

4.4.1 Item Separation Index

For simplicity, we assume that we are working with a set of items which show a good fit to the Rasch model. In this setting, we can obtain estimates of the item parameters $\hat{\beta}_i$, $i = 1, \ldots, I$, and the person parameters $\hat{\theta}_p$, $p = 1, \ldots, P$. Naturally, these estimates are not perfect, and we have standard errors for all estimates which we denote by $se(\hat{\beta}_i)$ and $se(\hat{\theta}_p)$. The following presentation is based on Wright and Masters (1982) and Andrich and Marais (2019).

We start with the presentation of the item separation index. We denote the variance of the item parameter estimates $\hat{\beta}_i$, calculated over all items, by SD_I^2:

$$SD_I^2 = \frac{\sum_{i=1}^{I}(\hat{\beta}_i - \bar{\beta})^2}{I - 1}.$$

Here, $\bar{\beta}$ denotes the mean of the item parameter estimates.[8]

We further denote the mean of the squared standard errors by

$$MSE_I = \frac{1}{I}\sum_{i=1}^{I} se(\hat{\beta}_i)^2.$$

We first calculate the following term, which can be regarded as an item parameter variance that is adjusted for the estimation error:

$$SA_I^2 = SD_I^2 - MSE_I.$$

[8]When we employ the constraint that the sum of all item parameters is zero, $\bar{\hat{\beta}}$ is zero as well. When we use the alternative constraint that the first item parameter is zero, this corresponds to an additive shift for all item parameters, but does not alter the location of the item parameters relative to each other. Therefore, the variance is the same, regardless of which constraint we use.

Second, we calculate the following term, which can be considered as an estimate of the overall root mean squared error of estimation[9]:

$$SE_I = \sqrt{MSE_I}.$$

The item separation index G_I is now calculated as

$$G_I = \frac{SA_I}{SE_I}.$$

The item separation index thus compares the standard deviation of the item parameters (i.e., how widely do the item parameters spread), which has been adjusted for the estimation error, with the root mean squared estimation error (i.e., how uncertain are the estimates of the item parameters). High values indicate a large value of SA_I, and a (comparatively) low value of SE_I. Such values thus indicate an item pool with accurately estimated item parameters that spread widely. Low values, on the other hand, indicate a small value of SA_I, and a (comparatively) large value of SE_I. This indicates that the item parameters have a low variance and tend to be uncertain estimates. Samples from populations with a high item separation index are therefore considered more suitable to separate between items of different difficulty levels.

4.4.2 Person Separation Index

By switching the role of items and persons, we can also obtain a person separation index. As before, we begin by defining SD^2_P as the variance of the person parameter estimates $\hat{\theta}_p$, calculated over all respondents. MSE_P denotes the mean of the squared standard errors $se(\hat{\theta}_p)^2$

$$MSE_P = \frac{1}{P}\sum_{p=1}^{P} se(\hat{\theta}_p)^2.$$

We now calculate

$$SA^2_P = SD^2_P - MSE_P$$

and

$$SE_P = \sqrt{MSE_P}.$$

This leads to the person separation index

$$G_P = \frac{SA_P}{SE_P}.$$

[9]Note that the estimation error for the item parameters is also termed calibration error in the literature.

In analogy to the item separation index, this term indicates how well respondents are separated by a set of test items. Higher values indicate that a test is more suited to differentiate between respondents of different ability levels.

The term $\frac{G_P^2}{1+G_P^2}$ is also considered an estimate of a test's reliability in Rasch measurement (Wright & Masters, 1982). As Andrich and Marais (2019) note, this term is conceptually related to analogous terms from classical test theory. It may take on values between 0 and 1, and values close to 1 indicate a high reliability. Tennant and Conaghan (2007), for example, write that it should be above 0.7 for group use and above 0.85 for individual use. The G_P index is available in `eRm`, as listed in Table 9.1.

4.5 Evaluation Through Model Comparisons

From the modeling point of view, when the items of a psychological test are not well described by the Rasch model, more flexible models are needed. In the following we describe models that mathematically contain the Rasch model as a special case. They relax some of its particularly strict assumptions from a modeling point of view – and thereby loose some of its particularly important properties from a measurement point of view. Please note that, while we present these models as extensions of the Rasch model in a mathematical sense, which we believe is helpful didactically, historically these models were developed independently and in parallel (cf., e.g., Embretson & Reise, 2000).

Later in this section we will see how we can judge by means of model comparison approaches which model is better suited for describing the responses to a given psychological test.

4.5.1 Models with Additional Item Parameters

Like the Rasch model, the models we describe here are suited for dichotomous items (for models with more than two response options, see Chapter 11). What makes these models more flexible than the Rasch model from a modeling point of view is that they have additional item parameters. This allows them to capture some of the patterns we have seen as violations of the Rasch model above, for example in Section 4.1.2.

4.5.1.1 Two-Parameter Model

The Rasch model assumes that every item has the same discrimination. Birnbaum's (1968) two-parameter logistic (2PL) model relaxes this assumption by allowing the ICCs to have different slopes. For every item, it introduces a new parameter α_i, called the slope or discrimination. The formula for the

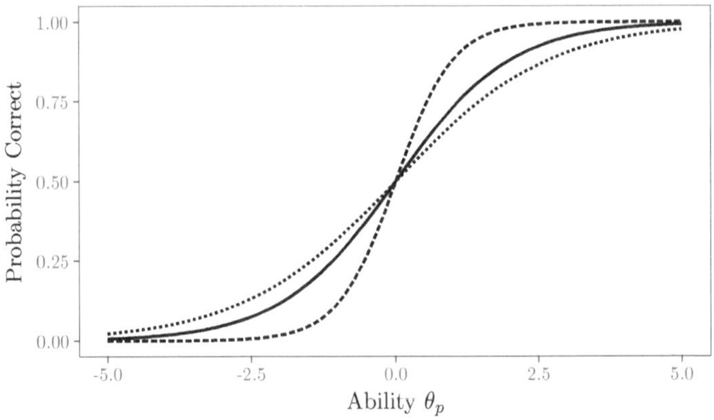

FIGURE 4.7: Illustration of the effect of the discrimination parameter α_i on the ICC. The discrimination parameter is $\alpha_i = 1$ for the solid line, $\alpha_i = 0.75$ for the dotted line, and $\alpha_i = 2$ for the dashed line.

2PL model is

$$\Pr(U_{pi} = 1 \mid \theta_p, \alpha_i, \beta_i) = \frac{\exp\{\alpha_i \cdot (\theta_p - \beta_i)\}}{1 + \exp\{\alpha_i \cdot (\theta_p - \beta_i)\}}. \tag{4.1}$$

The Rasch model is a special case of the 2PL in which the discrimination parameter $\alpha_i = 1$ for every item. When $\alpha_i > 1$, the ICC of item i will have a larger slope, so the item will do a better job of discriminating test takers. When $\alpha_i < 1$, the ICC of item i will have a smaller slope, so the item will do a worse job of discriminating test takers. Examples of 2PL ICCs are shown in Figure 4.7.

Because each item in the 2PL can have a different slope, the ICCs of different items can intersect. This means that specific objectivity (see Section 2.4.3) does not hold for the 2PL model. That said, allowing each item to have a different capacity for discriminating test takers may be more realistic than the Rasch model's strict requirement of equal discrimination.

In practice, deciding whether to use the Rasch or 2PL model also depends on your goals. When developing a new test, it may be worth searching for items that meet the more stringent requirements of the Rasch model, since the Rasch model is considered to have better properties than the 2PL from a measurement point of view. However, when dealing with the data from an existing test or questionnaire, from a modeling point of view the 2PL may be more appropriate for describing the response patterns in a particular data set.

Before moving on, it should also be noted that test takers' scores are no longer sufficient statistics for their ability in the 2PL model. This means that we can get different ability estimates for two people who have answered the same overall number of items correctly. Items with different discriminations are weighted differently, so usually only test takers who solve exactly the same items will have the same estimated ability.

The fact that the sum scores are no longer sufficient statistics means that we cannot use the conditional maximum likelihood approach to estimate the parameters of the 2PL. Accordingly, only those R packages using other estimation approaches can be used for estimating the 2PL as well as the following 3PL and 4PL models, as we will see in Part III of this book. This also confirms that it does not make sense to compare test takers using their sum scores without first checking whether the Rasch model provides a reasonable account of their behavior.

Finally, we would like to mention that sometimes the following variant of the equation of the 2PL model is used:

$$\Pr(U_{pi} = 1 \mid \theta_p, \tilde{\alpha}_i, \tilde{\beta}_i) = \frac{\exp\{\tilde{\alpha}_i \cdot \theta_p + \tilde{\beta}_i\}}{1 + \exp\{\tilde{\alpha}_i \cdot \theta_p + \tilde{\beta}_i\}}.$$

Here, the discrimination parameter α_i and the difficulty parameter β_i are replaced by a slope parameter $\tilde{\alpha}_i$ and an intercept parameter $\tilde{\beta}_i$. A comparison of both equations shows that $\tilde{\alpha}_i = \alpha_i$ and $\tilde{\beta}_i = -\beta_i \cdot \alpha_i$. The slope parameter can thus be interpreted in the same way as the discrimination parameter, whereas the intercept parameter increases with the discrimination and the easiness of an item.

4.5.1.2 Three-Parameter Model

Birnbaum's (1968) three-parameter logistic (3PL) model incorporates an additional parameter γ_i to account for guessing. Its ICC is given by

$$\Pr(U_{pi} = 1 \mid \theta_p, \alpha_i, \beta_i, \gamma_i) = \gamma_i + (1 - \gamma_i) \cdot \frac{\exp\{\alpha_i \cdot (\theta_p - \beta_i)\}}{1 + \exp\{\alpha_i \cdot (\theta_p - \beta_i)\}}.$$

The new parameter γ_i is the lower asymptote of the ICC. This means that the probability that any test taker solves the item is larger than γ_i. Even test takers with very low ability will solve the item with a probability of at least γ_i. The lower a test taker's ability, the closer the probability of solving the item will be to γ_i. This is demonstrated by Figure 4.8. When $\gamma_i = 0$, the 3PL reduces to the 2PL.

If we assume that $\gamma_i > 0$, this conceptually means that a test taker of very low (technically, infinitely low) ability will still have a positive probability of solving an item, which is a very sensible property when modeling multiple choice items. In multiple choice tests, examinees do not need to produce the correct response. Instead, they are provided with two or more possible answers and must simply select which of them is correct. For example, if there are three response alternatives, the probability that a test taker selects the correct response by chance is 1/3. In this case, it makes sense for the probability of a correct response to be at least 1/3, even for test takers having very low ability.

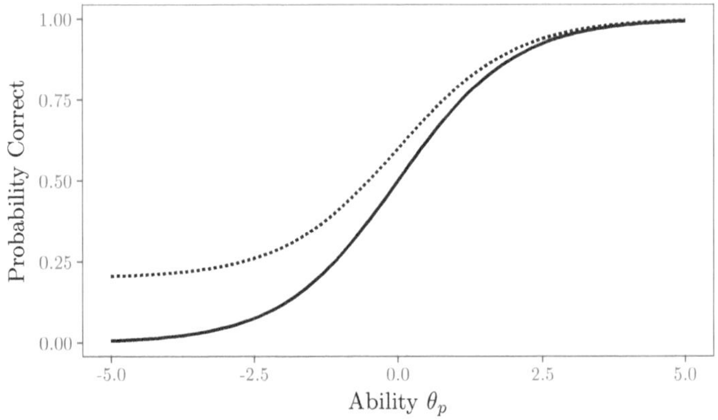

FIGURE 4.8: Illustration of the effect of the guessing parameter γ_i on the ICC. The guessing parameter $\gamma_i = 0$ for the solid line and $\gamma_i = 0.2$ for the dotted line.

In this example, we assumed that any of the three response alternatives is equally likely to be selected when the test taker does not know the answer. However, for real multiple choice items, it is possible that different response alternatives will be selected with different probabilities by examinees that do not know the correct response. Often, test designers will include response alternatives that are nearly correct in order to distract test takers who do not know the correct answer. These alternatives may be selected more than 1/3 of the time. Alternatively, some response alternatives are so obviously wrong that few test takers would be tempted to select them.

In either of these cases, the actual probability of correctly guessing the item may be different than 1/3. Thus, it would be more informative to allow γ_i to vary across items with the same number of response alternatives, rather than assuming it is one divided by the number of response alternatives. Unfortunately, to freely estimate γ_i for every item is only possible when the sample size is large. Test takers of very low ability are rare, but a number of them are needed to estimate the lower asymptote γ_i.

While often the same model is used for all items in a test, it is also possible to specify different IRT models for different items. For example, in a test that consists of a combination of multiple choice and open questions, also termed constructed response items, we could expect guessing to occur only in the multiple choice items. In this case it would be possible to use the 3PL model for the multiple choice items and the 2PL model for the constructed response items. The Trends in International Mathematics and Science Study (TIMSS), for example, since 1999 uses the 3PL model for multiple choice items, the 2PL model for constructed response items worth 1 point, and the Generalized Partial Credit model (an extension of the 2PL model for polytomous items, cf. Chapter 11) for constructed response items worth up to 2 points (von Davier, 2020).

As a side note, recall that the 2PL is the special case of the 3PL where $\gamma_i = 0$. We can get a different two parameter model by setting the discrimination parameter $\alpha_i = 1$ and allowing γ_i and β_i to vary. In this case, all of the items have the same discrimination, as in the Rasch model, but may have different guessing rates. This model is sometimes called the one-parameter logistic model with guessing or 1PL-G. This model, as well as the 2PL and 3PL models, can be estimated in R by means of MML estimation using either the `mirt` or `TAM` packages. They may also be estimated using Bayesian methods in Stan (or JAGS).

4.5.1.3 Four-Parameter Model

The 3PL model can be even further generalized to a four-parameter logistic (4PL) model (Barton & Lord, 1981). The ICC of this model is given by

$$\Pr(U_{pi} = 1 \mid \theta_p, \alpha_i, \beta_i, \gamma_i, \delta_i) = \gamma_i + (\delta_i - \gamma_i) \cdot \frac{\exp\{\alpha_i \cdot (\theta_p - \beta_i)\}}{1 + \exp\{\alpha_i \cdot (\theta_p - \beta_i)\}}.$$

It ensures that the upper asymptote of the ICC, that is, the asymptote for the probability of a correct response when θ_p increases, is not equal to 1 as in the 3PL model, but to δ_i. Therefore, even very proficient respondents may provide an incorrect response to a task with a probability of at least $1 - \delta_i$. Conceptually, the additional item parameter δ_i has the meaning of an inattention or slipping parameter: It provides an estimate how likely even extremely proficient respondents may provide an incorrect response to a given item. It follows from this presentation that the 3PL model is a special case of the 4PL model, which is obtained when δ_i is set to 1.

While the 4PL model seems plausible from a practical perspective, the parameter estimation is even more demanding than for the 3PL model. Whereas a large number of respondents of low ability is required to estimate the lower asymptote parameter γ_i, also many respondents of high ability are necessary to estimate the upper asymptote parameter δ_i. An alternative approach to working with this model consists of not estimating these parameters, but fixing them to specific values, such as 0.98 for δ_i. An example is the study of Rulison and Loken (2009), who used this value in the context of adaptive testing. In R this model can be estimated with `mirt` and, using Bayesian methods, with Stan (or JAGS).

4.5.1.4 Sample Size Requirements

Like for the Rasch model, several studies have addressed the question of what size of the calibration sample is necessary to obtain an acceptable accuracy of the item parameter estimates for the models presented above. We briefly summarize the literature here, which is reviewed more extensively by De Ayala

(2009). As was already stated in Section 3.7, R users can carry out similar studies themselves.

As was already the case for the Rasch model (cf. Section 3.7), these guidelines should be seen merely as rough indicators, which sample sizes could be practically useful. The accuracy of the item parameter estimates does not only depend on the sample size, but also on other factors, such as the number of items and the alignment and shape of the item and person parameter distributions.

Depending on the conditions of measurements, sample sizes of 200 (Drasgow, 1989) or 500 (Stone, 1992) (for 5 to 20 items, respectively) could be enough to obtain sufficiently accurate estimations for the 2PL model. For the 3PL model, a sample size of 1000 respondents was found to be adequate by Yen (1987), which is also suggested as a minimum by De Ayala (2009) (for 20 items and under favorable conditions) in his review of the literature.

As already explained in Section 4.5.1.2, estimating the lower asymptote of the 3PL may be particularly difficult. Accordingly, De Ayala (2009) also mentions that fixing the guessing parameters to a reasonable non-zero value may help when estimating them freely proves impossible for a given sample.

So far, no literature seems to be available that provides guidelines for sample size requirements for the 4PL model, but it can be expected that typically even larger samples are necessary for estimating this model than for the 3PL model. Indications that the employed sample size was not sufficient for reliably estimating the model parameters are convergence problems and large standard errors.

4.5.2 Likelihood Ratio Tests

When different IRT models are available to describe the psychological test data, from a modeling point of view this leads to the practical question which of these models should be chosen.

One possible solution to this problem is the application of likelihood ratio tests. These tests can be used to choose between two models if these models meet the following three assumptions: First, the two models have to be nested, that is, one model is obtained by fixing some of the parameters of the other model to specific values. For instance, likelihood ratio tests can be used for choosing between the 2PL model and the Rasch or 1PL model.[10] These two models are nested because the Rasch model is obtained by fixing all α_i parameters of the 2PL model to 1, as was discussed in Section 4.5.1.1. By means of comparing the two models we can statistically test whether the additional discrimination parameters of the 2PL are actually necessary to describe the

[10] In the IRT literature, the one-parameter logistic or 1PL model is formulated with equal slopes for all items, but without fixing the slope to 1. The Rasch model and the 1PL model come from different philosophies, which we have termed the measurement and the modeling point of view. However, mathematically they are equivalent and the specific value of the slope parameter is absorbed into the metric for the latent continuum (De Ayala, 2009).

data, or whether the discrimination parameters are so similar that the Rasch model describes the data just as well.

The second, less well known prerequisite for the likelihood ratio test is that the fixed parameters in the more restrictive model must not lie on the boundary of the parameter space of the more general model. For instance, the 2PL model can be obtained from the 3PL by fixing the values of the γ_i parameters to 0, as was discussed in Section 4.5.1.2. However, in the more general 3PL these parameters can only take on values between 0 and 1, so fixing them to 0 corresponds exactly to the lower boundary of their parameter space. In such cases, the likelihood ratio test should not be applied. Brown, Templin, and Cohen (2015) indicate deflated Type I error rates for comparable scenarios, and Chalmers, Pek, and Liu (2017) found similar problems for the computation of confidence intervals.

The third assumption is that at least the more general model is the true model that accurately describes the underlying item response pattern. This assumption cannot be tested, therefore both models should offer plausible descriptions of the data. Note that a model comparison via information indices does not make this assumption.

To apply a likelihood ratio test, first we estimate the item parameters under both models using a maximum likelihood approach. This leads to two sets of item parameter estimates, $\hat{\beta}^{(1)}$ for the first, more restrictive model, and $\hat{\beta}^{(2)}$ for the second, more general model. The null hypothesis of the likelihood ratio test is that the more restrictive model is true. If this is the case, the parameter estimates of the more general model should be very close to those of the more restrictive model. Furthermore, the likelihoods of both models should be close to each other.

Using this idea, we can now obtain the test statistic of the likelihood ratio test by considering the marginal log-likelihoods of both models, $\log(L_{\boldsymbol{u}}(\hat{\boldsymbol{\beta}}^{(1)}))$ and $\log(L_{\boldsymbol{u}}(\hat{\boldsymbol{\beta}}^{(2)}))$, as:

$$LR = -2\log\left(\frac{(L_{\boldsymbol{u}}(\hat{\boldsymbol{\beta}}^{(1)}))}{(L_{\boldsymbol{u}}(\hat{\boldsymbol{\beta}}^{(2)}))}\right).$$

Under the null hypothesis, and given the three assumptions stated above, this statistic follows a χ^2 distribution. Its degrees of freedom correspond to the number of restricted parameters. This test is available via the command `anova` in R and its application will be illustrated in Chapter 7.

4.5.3 Information Criteria

Another important approach to select among several plausible models is based on information criteria, such as the Akaike Information Criterion (AIC) or the Bayesian Information Criterion (BIC, also termed Schwarz Information Criterion). Computationally, these approaches lead to numbers that are compared for both models. In contrast to the likelihood ratio test, these criteria do not assume that one of the two models is true, or that one of the models is a restricted version of the other one (i.e., that the models are nested). Drawbacks of information criteria are that their comparison does not lead to a statistical test and that different information criteria might come to contradicting conclusions.

Of the two criteria, the BIC tends to select smaller models with fewer parameters. This is clear when we look at the formulas for computing AIC and BIC:

$$AIC = -2 \cdot \log L_{\boldsymbol{u}}(\hat{\boldsymbol{\beta}}) + 2 \cdot m,$$

where $\log L_{\boldsymbol{u}}(\hat{\boldsymbol{\beta}})$ is the log-likelihood evaluated at the ML-estimates $\hat{\boldsymbol{\beta}}$ for the data $\boldsymbol{u}$ and m is the number of parameters of the model. In this formula, the log likelihood expresses how well the model fits the data. Due to the factor -2, the *smaller* AIC the better. The number of parameters represents the model complexity. The complexity is penalized[11], because otherwise the most flexible model with the most parameters would always win the comparison.

When we compare the formula for the BIC

$$BIC = -2 \cdot \log L_{\boldsymbol{u}}(\hat{\boldsymbol{\beta}}) + \log(P) \cdot m$$

to that of the AIC, we find that for the BIC the number of parameters m is penalized with a factor of $\log(P)$, where P is the sample size. It can be shown that the BIC penalizes the number of parameters more strictly as soon as the sample size is greater than eight[12], which applies to any sample size that is reasonable for estimating the models we consider here. Because of this different penalization factor, the AIC and BIC criteria do not always agree in model comparisons, with the BIC favoring more sparse models.

Many R packages, such as `mirt`, offer sample-adjusted variants of these information indices, such as the corrected Akaike information criterion (AICc; Hurvich & Tsai, 1993; Sugiura, 1978) and the sample-size adjusted Bayesian information criterion (SABIC; Sclove, 1987). In practical analyses, they are interpreted in the same way as AIC and BIC.

4.6 Exercises

1. Figure 4.9 (left) shows the result of a graphical test. Determine which item(s) exhibit DIF.
2. Figure 4.9 (right) shows the result of a graphical test for a different data set. Can we conclude that in this example all items exhibit DIF?

[11]When we add 2× the number of parameters m to $-2 \cdot \log L_{\boldsymbol{u}}(\hat{\boldsymbol{\beta}})$, the value of the AIC goes up. Information criteria follow the principle "the smaller the better", so any quantity that increases their value works as a penalty.

[12]Because the factor $\log(P)$ is greater than the factor 2 for all $P > 8$.

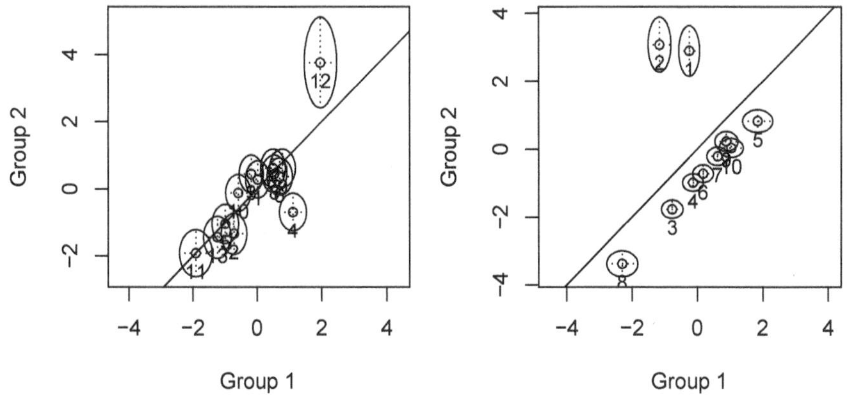

FIGURE 4.9: Figures used in the end-of-chapter exercises.

3. Which other test could we use to test individual items for DIF, and why do we have to select anchor items for any item-wise test?
4. Show mathematically that DIF leads to a violation of the multiplication condition for specific objectivity (cf. Section 2.4.3, for example in logit form on p. 30, and Section 4.2).
5. Briefly explain the logic underlying a χ^2 goodness-of-fit test.
6. What are Rasch residuals? Name two examples of statistics that are calculated based on Rasch residuals.
7. What is the principle behind the Q_3 statistic?
8. Which goodness-of-fit statistics for the Rasch model have been suggested for small samples?

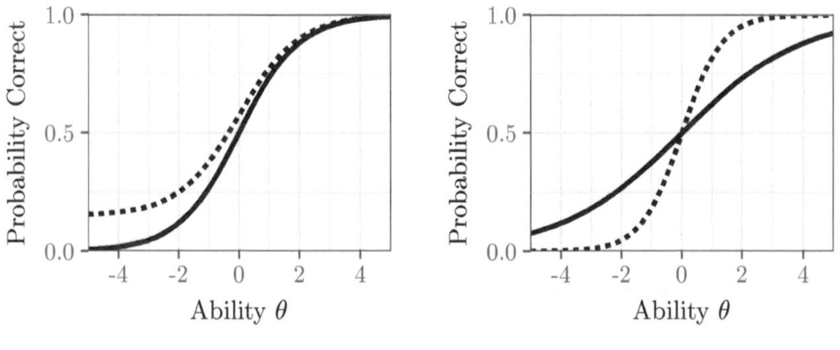

FIGURE 4.10: Figures used in the end-of-chapter exercises.

9. Determine which of the ICCs in Figure 4.10 (left) corresponds to the 2PL model and which corresponds to the 3PL model.
10. Determine which of the items in Figure 4.10 (right) has the larger discrimination parameter.
11. Name two approaches for comparing models and their respective limitations.

Part II

Applications

5

Basic R Usage

CONTENTS

This second part of the book will present case studies demonstrating how the Rasch model can be estimated, evaluated and used to understand real test data. In this chapter, we introduce some R basics that will be useful for going through these examples. If you are already familiar with R, you can skip this chapter.

Section 5.1 explains how to install on your computer both the R software itself and additional R packages, which extend the functionality of R. Section 5.2 shortly introduces RStudio, a widely used code editor and semi-graphical user interface for R. Section 5.3 shows how to load or read in data sets. Sections 5.4 and 5.5 explain important fundamentals of working with R objects by summarizing and extracting information from data matrices and working with lists. If you have no previous experience with R, we recommend first trying out the commands in this chapter and then working through the exercises in Section 5.6 before moving on to the following chapters.

5.1 Installation of R and Add-On Packages

To install the latest version of R, go to `https://www.cran.r-project.org/`, choose the download option corresponding to your operating system and follow the instructions.

After first installing R, a collection of functions for basic statistical analyses is already included. For specialized analyses, like the ones we want to perform in the following chapters, additional packages must be installed. The rationale behind this is that not every R user wants to do psychometric analyses, and those who do not have no need for packages that are specific to this field. Therefore, the default R installation includes only functions that most users will want to use, but literally thousands of add-on packages are available to extend the functionality.[1]

Before using a new package for the first time, it needs to be installed using the command `install.packages()`. For example,

```
> install.packages("TAM")
```

installs the `TAM` package on the computer. In addition, every time we want to use this package in a new R session, we need to activate its contents by loading the package (from a directory referred to as a library) with the `library()` command. For example,

```
> library("TAM")
```

loads the `TAM` package.[2] To view a list of all contents provided by a package, use the `help` argument to `library()`. For example, typing

```
> library(help = "TAM")
```

will list all functions and data provided by the package `TAM`.

Package developers are required to document the functionality provided by their package. This documentation is provided in the reference manual, that is available, e.g., as one joint document on the package's webpage at the Comprehensive R Archive Network CRAN. For the `TAM` package, for example, the URL is `https://cran.r-project.org/package=TAM`. Information about each individual function or data set is also available in a help file.

In the case of functions, the help file documents what the function does, including its input arguments and output values. In the case of data sets, the help file describes the structure and content of the data. Help files can be viewed using the `help()` function. To view the help file for the `data.fims.Aus.Jpn.scored` data set in the `TAM` package, for instance, we could type

```
> help("data.fims.Aus.Jpn.scored")
```

[1]To get an overview over packages available for different areas of statistics, so-called task views are available at `https://cran.r-project.org/web/views/`, including one on psychometrics.

[2]The `library()` command also works without quotes, but we use them for consistency.

or, alternatively, we could place the ? operator before the name of the function or data set

```
> ?data.fims.Aus.Jpn.scored
```

to open the help file.

Some packages (such as **eRm** and **mirt**) also provide additional documentation in the form of vignettes, which are available, e.g., via their CRAN pages and explain in detail how to use certain functions.

5.2 Code Editors and RStudio

When typing commands in a programming language like R, a single typo or missing symbol can result in non-executable code. This is why it is recommended to use a code editor with syntax highlighting to write R commands. Syntax highlighting makes it easier to read and check the code, e.g., by means of highlighting correct code chunks of a certain type in a certain color. Many code editors with syntax highlighting for R are available for all operating systems. After typing and checking the commands in the editor, they can be pasted into the R console to be executed.

An R code editor that has become particularly popular because it offers additional amenities is RStudio. It provides not only a code editor with syntax highlighting and auto-completion, but also a graphical user interface for file handling etc.

To install RStudio, go to `https://rstudio.org/download/desktop` (after first installing R). Choose a license for RStudio Desktop (not RStudio Server) according to your operating system.

The RStudio user interface (see Figure 5.1) is divided into four cells for R scripts (top left), that can be executed in the R console with the "Run" button, the R console itself (bottom left), the workspace (top right), where all active objects are listed, and a panel where help files and plots are displayed (bottom right). To create a new R script, use "File – New File" in the top menu; to open an existing R script, use "File – Open File". By convention R scripts have the file extension `.R`.

An important aspect of working with R is that by saving the R code for all steps of the analysis—rather than conducting an analysis by clicking on the buttons of a purely graphical user interface and typically not remembering how it was done exactly a few days later—the analysis is fully reproducible. For this reason, we strongly recommend conducting the entire analysis, including any data pre-processing steps, by means of commands in the R script, and never making any changes in the data manually.

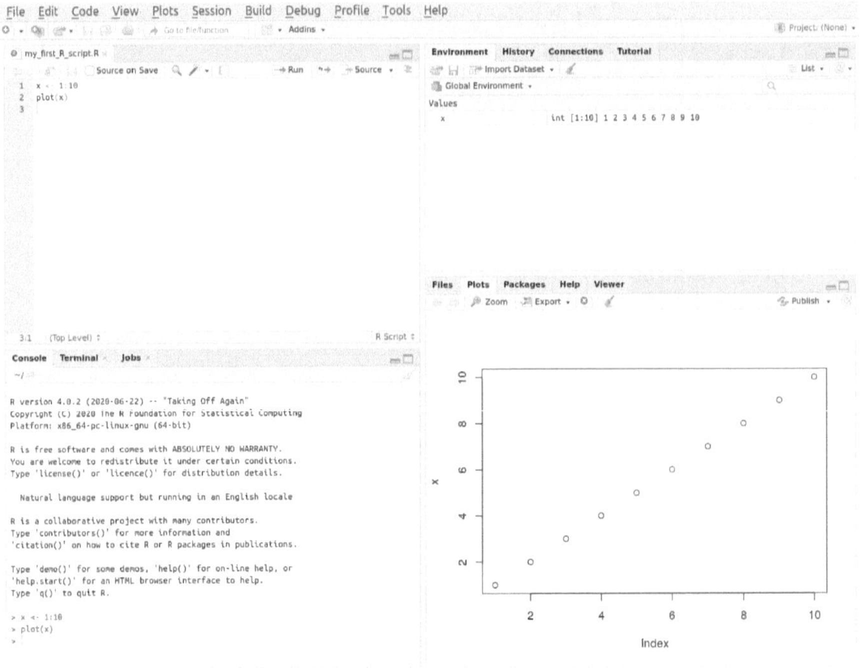

FIGURE 5.1: Screenshot of the RStudio user interface.

When the R script contains all commands for data pre-processing, analysis, generation of plots and tables etc., this script together with the original data set makes the entire analysis reproducible.[3] When you stick to this rule, it is not necessary to save all intermediate objects or results created during an R session. This means that it is safe to select "Don't Save" when, upon closing RStudio, you are asked whether you want to save the workspace image. It only makes sense to permanently store results in a file when they take a long time to compute. This can be done explicitly by means of including a `save()` command in the R script.

The use of R for reproducible research, including automated report generation, is discussed in more detail in the book by Gandrud (2014).

5.3 Loading and Importing Data

We will use the `data.fims.Aus.Jpn.scored` data set from the `TAM` package for the case studies in this chapter. The `data.fims.Aus.Jpn.scored` data set

[3]For certain analyses, that involve random number generation, so called seeds need to be set additionally for full reproducibility.

contains scored responses for a subset of items from Australian and Japanese students in the First International Mathematics Study (FIMS, Husén, 1967). After `TAM` has been installed and loaded, we can make the data set visible and save it into a new object named **`fims`** (just for brevity) by entering

```
> data(data.fims.Aus.Jpn.scored, package = "TAM")
> fims <- data.fims.Aus.Jpn.scored
```

The **`data.fims.Aus.Jpn.scored`** data set has already been made available by loading the `TAM` package. If you want to use your own data, this will typically require loading/reading data from one or more files into the R session. R is capable of loading/reading in a number of different types of data, including spreadsheets and SPSS data files. Examples include the functions **`load()`** for data files stored from within R (file extensions **`.Rdata`** or **`.rda`**), **`read.csv()`** and **`read.xlsx()`** (the latter requiring the add-on package **`xlsx`**) for data files stored from spreadsheets (file extensions **`.csv`**, for comma separated values, or **`.xlsx`**), or **`read.spss()`** (requiring the add-on package **`foreign`**), for data files stored from SPSS (file extension **`.sav`**).

For a more comprehensive introduction to loading and manipulating data in R, see Spector (2008).

5.4 Getting Information About Persons and Variables

In this section we will introduce different commands for inspecting and accessing different parts of a data set. These are necessary for getting an overview over a new dataset, but also, for example, for selecting certain observations or variables in a data set for further analyses.

We can gain general information on the structure of a data set by using

```
> str(fims)
'data.frame': 6371 obs. of  16 variables:
 $ SEX    : int  1 1 1 1 1 1 1 1 1 1 ...
 $ M1PTI1 : num  1 0 1 1 1 1 0 1 0 0 ...
 $ M1PTI2 : num  0 1 0 1 1 1 0 1 0 1 ...
 $ M1PTI3 : num  1 1 1 1 1 1 1 0 1 1 ...
 $ M1PTI6 : num  1 0 0 1 1 0 0 1 0 0 ...
 $ M1PTI7 : num  0 0 0 0 0 0 0 0 0 0 ...
 $ M1PTI11: num  1 0 0 1 1 0 1 1 0 1 ...
 $ M1PTI12: num  0 0 0 1 0 0 1 0 0 0 ...
 $ M1PTI14: num  0 1 0 0 1 0 1 1 1 0 ...
 $ M1PTI17: num  1 0 0 0 0 0 1 0 0 0 ...
 $ M1PTI18: num  0 0 1 0 1 1 0 1 0 1 ...
 $ M1PTI19: num  0 0 0 0 0 0 0 0 0 0 ...
```

```
 $ M1PTI21: num  0 0 0 0 0 1 0 0 0 0 ...
 $ M1PTI22: num  0 0 0 0 0 0 0 0 1 1 ...
 $ M1PTI23: num  1 1 1 0 1 0 0 1 0 1 ...
 $ country: int  1 1 1 1 1 1 1 1 1 1 ...
```

whose output indicates that `fims` is a `data.frame` with 6371 observations of 16 variables of different types. For each variable, the values for the first ten persons are displayed. For example, we can see that the first variable is named `SEX` and that its first ten observations are all 1. We can also see the data type for each variable. For instance, `int` denotes an integer variable (i.e., a variable containing whole numbers). As we can see, the variable `SEX` is encoded as an integer variable, which is not ideal. We will come back to this aspect below.

The following variables represent the mathematics items. We notice that they are not numbered consecutively, because only a subset of items is stored in this data set. For example, there is no item called "M1PTI4" or "M1PTI5", so that "M1PTI6" is not the sixth but the fourth item listed. Throughout this and the following chapters, we will have to keep this in mind when addressing the items.

The interested reader can find the items, including the subset from the "Test A in Mathematics" used here, together with further information about the FIMS study at:

`https://www.gu.se/en/center-for-comparative-analysis-of-educational-achievement-compeat/studies-before-1995/first-international-mathematics-study-1964`

A different function that can help us get an impression of the data set is the `head()` function, which displays the first few rows of the data set. In every data matrix, each observation corresponds to a row, and each variable to a column. For this example

```
> head(fims)
  SEX M1PTI1 M1PTI2 M1PTI3 M1PTI6 M1PTI7 M1PTI11 M1PTI12 M1PTI14
1   1      1      0      1      1      0       1       0       0
2   1      0      1      1      0      0       0       0       1
3   1      1      0      1      0      0       0       0       0
4   1      1      1      1      1      0       1       1       0
5   1      1      1      1      1      0       1       0       1
6   1      1      1      1      0      0       0       0       0
  M1PTI17 M1PTI18 M1PTI19 M1PTI21 M1PTI22 M1PTI23 country
1       1       0       0       0       0       1       1
2       0       0       0       0       0       1       1
3       0       1       0       0       0       1       1
4       0       0       0       0       0       0       1
5       0       1       0       0       0       1       1
6       0       1       0       1       0       0       1
```

we see again the first few values for the variables `SEX`, `country` and the mathematics test items.

If we are interested in a summary of descriptive statistics for the `fims` data set, we can use the `summary()` function. The `summary()` function summarizes each column of a `data.frame` in a way that is consistent with its type. For numeric data, summary shows a column's mean, minimum, and maximum along with certain quantiles.

To view a summary of `fims`, we can type

```
> summary(fims)
      SEX             M1PTI1            M1PTI2            M1PTI3
 Min.   :1.000   Min.   :0.0000   Min.   :0.0000   Min.   :0.0000
 1st Qu.:1.000   1st Qu.:1.0000   1st Qu.:1.0000   1st Qu.:1.0000
 Median :1.000   Median :1.0000   Median :1.0000   Median :1.0000
 Mean   :1.479   Mean   :0.7732   Mean   :0.7616   Mean   :0.8465
 3rd Qu.:2.000   3rd Qu.:1.0000   3rd Qu.:1.0000   3rd Qu.:1.0000
 Max.   :2.000   Max.   :1.0000   Max.   :1.0000   Max.   :1.0000
     M1PTI6            M1PTI7            M1PTI11           M1PTI12
 Min.   :0.0000   Min.   :0.0000   Min.   :0.0000   Min.   :0.000
 1st Qu.:0.0000   1st Qu.:0.0000   1st Qu.:1.0000   1st Qu.:0.000
 Median :1.0000   Median :0.0000   Median :1.0000   Median :0.000
 Mean   :0.5712   Mean   :0.1631   Mean   :0.7963   Mean   :0.342
 3rd Qu.:1.0000   3rd Qu.:0.0000   3rd Qu.:1.0000   3rd Qu.:1.000
 Max.   :1.0000   Max.   :1.0000   Max.   :1.0000   Max.   :1.000
    M1PTI14           M1PTI17           M1PTI18           M1PTI19
 Min.   :0.0000   Min.   :0.0000   Min.   :0.0000   Min.   :0.000
 1st Qu.:0.0000   1st Qu.:0.0000   1st Qu.:0.0000   1st Qu.:0.000
 Median :0.0000   Median :0.0000   Median :1.0000   Median :0.000
 Mean   :0.3998   Mean   :0.2593   Mean   :0.6111   Mean   :0.204
 3rd Qu.:1.0000   3rd Qu.:1.0000   3rd Qu.:1.0000   3rd Qu.:0.000
 Max.   :1.0000   Max.   :1.0000   Max.   :1.0000   Max.   :1.000
    M1PTI21          M1PTI22           M1PTI23          country
 Min.   :0.000   Min.   :0.0000   Min.   :0.0000   Min.   :1.000
 1st Qu.:0.000   1st Qu.:0.0000   1st Qu.:0.0000   1st Qu.:1.000
 Median :0.000   Median :0.0000   Median :1.0000   Median :1.000
 Mean   :0.235   Mean   :0.1868   Mean   :0.6936   Mean   :1.322
 3rd Qu.:0.000   3rd Qu.:0.0000   3rd Qu.:1.0000   3rd Qu.:2.000
 Max.   :1.000   Max.   :1.0000   Max.   :1.0000   Max.   :2.000
```

This output shows the minimum (`Min.`), the maximum (`Max.`), the mean, the median, and the first and third quartiles (`1st Qu.`, `3rd Qu.`) for each variable.

For the categorical variables `SEX` and `country`, this output is not ideal. Thus, we recode these two variables into unordered factors by means of the `as.factor()` command (for the `$` operator see below). From the `TAM` help page, we know that **SEX** encodes the test takers' gender with the levels 1 for male and 2 for female (which were the only genders recorded for this data set). We can label them accordingly by assigning these two strings to the two levels. The command `c()` is used in R to combine several elements into a vector or list – in this case the vector of the two labels for the two factor levels. The same is done for `country`.

```
> fims$SEX <- as.factor(fims$SEX)
> levels(fims$SEX) <- c("male", "female")
> fims$country <- as.factor(fims$country)
> levels(fims$country) <- c("Australia", "Japan")
```

When we repeat the **head()** and **summary()** commands

```
> head(fims)
   SEX M1PTI1 M1PTI2 M1PTI3 M1PTI6 M1PTI7 M1PTI11 M1PTI12
1 male      1      0      1      1      0       1       0
2 male      0      1      1      0      0       0       0
3 male      1      0      1      0      0       0       0
4 male      1      1      1      1      0       1       1
5 male      1      1      1      1      0       1       0
6 male      1      1      1      0      0       0       0
  M1PTI14 M1PTI17 M1PTI18 M1PTI19 M1PTI21 M1PTI22 M1PTI23
1       0       1       0       0       0       0       1
2       1       0       0       0       0       0       1
3       0       0       1       0       0       0       1
4       0       0       0       0       0       0       0
5       1       0       1       0       0       0       1
6       0       0       1       0       1       0       0
    country
1 Australia
2 Australia
3 Australia
4 Australia
5 Australia
6 Australia


> summary(fims[ , c("SEX", "country")])
     SEX              country
 male  :3319    Australia:4320
 female:3052    Japan    :2051
```

(the summary is shown only for the two recoded variables to save space; for the `[` operator see below), we find that the interpretation is clearer now: 3052 of the 6371 test takers are female, 4320 are Australian.

We can also gain information about the data matrix `fims` by looking at individual elements and submatrices. We access individual elements of `fims` in a variety of ways, one of which is to treat the `data.frame` as if it were a matrix. For example, we can determine whether the fifth test taker correctly answered the first item (named `M1PTI1`) by entering

```
> fims[5, "M1PTI1"]
[1] 1
```

or by remembering from above that `M1PTI1` is the second variable in the data set and using the equivalent command

```
> fims[5, 2]
[1] 1
```

for extracting the entry in the fifth row and second column of the data set.

We can also display the entries for the first five rows of the first five columns by entering

```
> fims[1:5, 1:5]
   SEX M1PTI1 M1PTI2 M1PTI3 M1PTI6
1 male      1      0      1      1
2 male      0      1      1      0
3 male      1      0      1      0
4 male      1      1      1      1
5 male      1      1      1      1
```

which shows the gender and the responses to the first four test items for each of the first five test takers. If we are not interested in seeing the gender of the test takers, we can focus on the responses to the first four test items by entering

```
> fims[1:5, 2:5]
  M1PTI1 M1PTI2 M1PTI3 M1PTI6
1      1      0      1      1
2      0      1      1      0
3      1      0      1      0
4      1      1      1      1
5      1      1      1      1
```

We can also get entire rows or columns of a matrix or `data.frame`. To get the entire first row of `fims`, which corresponds to all entries for the first test taker, we can type

```
> fims[1, ]
   SEX M1PTI1 M1PTI2 M1PTI3 M1PTI6 M1PTI7 M1PTI11 M1PTI12
1 male      1      0      1      1      0       1       0
  M1PTI14 M1PTI17 M1PTI18 M1PTI19 M1PTI21 M1PTI22 M1PTI23
1       0       1       0       0       0       0       1
    country
1 Australia
```

and to get the entire third column, which corresponds to all responses to the second item `M1PTI2`, we can type

```
> fims[, 3]
```

or, alternatively, we can use the `$` operator together with the variable name to address an individual variable

```
> fims$M1PTI2
```

(for these two examples we are not printing the output here to save space – they contain 6371 entries each).

Often we are also interested in the number of rows and columns of a `data.frame`. For item response data, the number of rows will usually correspond to the number of people who took the test. The number of columns will usually correspond to the number of test items, though it does not in this case, because the `fims` data set has two columns of demographic information. We can retrieve the number of rows and columns using the functions `nrow()` and `ncol()`, respectively. For example,

```
> nrow(fims)
[1] 6371
> ncol(fims)
[1] 16
```

and we can also get both the number of rows and the number of columns at the same time using the `dim()` function

```
> dim(fims)
[1] 6371   16
```

5.5 Addressing Elements in Lists

A `data.frame` and a `matrix` are two commonly used types of R objects. Another commonly used type is a list, which is essentially a collection of R objects. Collecting multiple R objects into a single place is useful, because it provides a way to group related objects. Lists are important to understand, because they are the output of many statistical functions in R.

We can create a list using the `list()` function. Suppose that we want to store the year in which the FIMS data were collected along with the data themselves. We could use a list to do this. For example,

```
> fims_dated <- list(data = fims, year_collected = "1964")
```

creates a list with two elements: `data` and `year_collected`. The value of the `data` entry is a copy of the FIMS data set. The value of the `year_collected` entry is the string "1964".

We can get the individual elements of this list using either the `$` or `[[` operators. Either of the following will get the year in which the FIMS data was collected

```
> fims_dated$year_collected
[1] "1964"
> fims_dated[["year_collected"]]
[1] "1964"
> fims_dated[[2]]
[1] "1964"
```

It turns out that we can access the columns of a `data.frame` in the same way, because a `data.frame` is a special type of list in which each column or variable is a list element. This means that every element of the list is a vector whose length is the number of rows in the `data.frame`. Thus, being a `data.frame`, the `fims` object is also a list whose elements are its columns, `SEX`, `M1PTI1`, etc. This allows us to access the `M1PTI1` column of `fims` by typing

```
> fims$M1PTI1
```

```
> fims[["M1PTI1"]]
```

or

```
> fims[[2]]
```

(again, the output of these functions has been omitted to save space).

R also allows for nested indexing. For example, we can get the first six rows of the `SEX` column of the `data` element of `fims_dated` by typing

```
> fims_dated$data$SEX[1:6]
[1] male male male male male male
Levels: male female
```

where the first `$` gets the `data` element from `fims_dated`, and the second gets the `SEX` column from that element. Finally, the `[1:6]` gets elements one through six of the `SEX` column of the `data` element.

5.6 Exercises

1. After installing R (and if you like RStudio), install the `psychotree` package and load it into the workspace.
2. Bring the data set `SPISA` from the `psychotree` package into the workspace.
3. Use the `str()` function to get information about the data set. What are its elements and their types?
4. The `SPISA` data set contains item responses from a general knowledge quiz and several covariates. The item responses are stored in the matrix `SPISA$spisa`. Get the number of test takers and test items.
5. Get the response of the 18th test taker to the 12th item.
6. Get the responses of test takers 65 through 70 to item 10.
7. Get all of test taker 422's responses. How many responses are there for each test taker?
8. Get the responses for items 37 through 45. These are the responses to the natural science items. Store them under the name `natSci_responses`.
9. Open the help file for the `rowSums()` function. Describe what it does.
10. Use the `rowSums()` function to compute the sum scores for each test taker in `natSci_responses` and store them under the name `scores`.
11. What is the highest number of items that test takers answered correctly?

6

R Package `eRm`

CONTENTS

The `eRm` package (Mair et al., 2021) contains functions for fitting the Rasch model as well as related models (in particular linear-logistic test models, cf. Section 10.1, and polytomous Rasch models, cf. Section 11.1 and Section 11.2) by means of conditional maximum likelihood estimation. In the following, we will discuss the practical application of the theory described in Chapters 3 and 4 for the Rasch model with `eRm`.

We will focus on those parts of the R commands and output that directly correspond to the theory part of the book. If you are interested in additional options or parts of the output not explained here, please refer to the documentation of the respective R package.

You will find this chapter on `eRm` and the following chapter on `mirt` to be the most elaborate ones. In order to avoid too many repetitions, we will not illustrate every package in the same detail. However, an overview of which estimation and evaluation approaches are available in which package(s) is provided in Table 9.1 starting on p. 192.

To install the eRm package before first using it, type

```
> install.packages("eRm")
```

Then, you can load its functionality into the workspace by typing

```
> library("eRm")
```

6.1 Item Parameter Estimation

We will fit a Rasch model for the first 400 test takers in the FIMS data set as an example. This data set contains scored responses from Australian and Japanese students in the First International Mathematics Study (Husén, 1967). It was already used in Chapter 5 and will be used for illustration in most of the R applications.

Before fitting the Rasch model, we need to separate the test responses in the FIMS data from the demographic information. As we have seen in Chapter 5, in the FIMS data the item responses are contained in columns 2-15, while demographic information is contained in columns 1 and 16. We would like to store the item responses and the gender information for the first 400 test takers in two variables. (We will not save the country information, because all of the first 400 test takers are from Australia.) One way of doing this[1] is by entering

```
> data(data.fims.Aus.Jpn.scored, package = "TAM")
> people <- 1 : 400
> responses <- data.fims.Aus.Jpn.scored[people, 2:15]
> gender <- as.factor(data.fims.Aus.Jpn.scored$SEX[people])
> levels(gender) <- c("male", "female")
```

explained in Chapter 5, the levels of SEX are 1 for male and 2 for female (which were the only genders recorded for this data set). In order to make many of the following outputs and plots more readable, we will abbreviate the item

[1]In the following code we will create two separate objects for responses and gender. This will make it easier to read some of the R commands below. Another option would be to merge them into a data frame. In that case, we would have to refer to the individual entries by means of the $ operator (cf. Section 5.5), for example in the form data_frame_name$responses. Many other R functions, such as lm() for linear models, allow the user to bypass the $ operator by stating the name of the data frame as the data argument. Then the individual entries can be addressed directly via a formula interface. However, for most of the analyses we are presenting in this book, the item response matrix plays the most central role, and in many cases will be the only part of the data that we want to hand over to the R functions.

names, which correspond to the column names of the item response matrix. They all start with the sequence "M1PTI", which we will substitute by the shorter "I" for item by typing

```
> colnames(responses) <- gsub("M1PTI", "I", colnames(responses))
```

We can then fit the Rasch model by entering

```
> rm_sum0 <- RM(responses)
```

where we use the name `rm_sum0` to indicate that the model was fit using the sum zero restriction, see also below.

From the presentation of the Rasch model in this and most other textbooks we are used to item parameters describing the difficulty of an item. The inverse of the item difficulty is the item easiness. The `eRm` package allows to extract both types of parameters, difficulty and easiness, because for some of the models in `eRm` easiness parameters are more convenient (Koller, Alexandrowicz, & Hatzinger, 2012). We will now show how to access both types of parameters in R, but also how the constraint for fixing the origin of the latent scale discussed in Section 3.2 shows in the R output.

To view the estimated item parameters, type

```
> summary(rm_sum0)

Results of RM estimation:

Call:  RM(X = responses)

Conditional log-likelihood: -1886.529
Number of iterations: 23
Number of parameters: 13

Item (Category) Difficulty Parameters (eta): with 0.95 CI:
    Estimate Std. Error lower CI upper CI
I2    -1.420      0.121   -1.658   -1.183
I3    -2.210      0.145   -2.494   -1.926
I6    -0.215      0.108   -0.426   -0.004
I7     2.364      0.170    2.031    2.697
I11   -1.420      0.121   -1.658   -1.183
I12    0.642      0.113    0.422    0.863
I14   -0.663      0.110   -0.879   -0.448
I17    1.152      0.122    0.913    1.391
I18   -0.565      0.109   -0.778   -0.351
I19    1.889      0.146    1.602    2.175
I21    1.578      0.134    1.315    1.841
I22    2.244      0.163    1.925    2.564
I23   -2.103      0.141   -2.379   -1.827
```

```
Item Easiness Parameters (beta) with 0.95 CI:
         Estimate Std. Error lower CI upper CI
beta I1     1.273      0.118    1.041    1.504
beta I2     1.420      0.121    1.183    1.658
beta I3     2.210      0.145    1.926    2.494
beta I6     0.215      0.108    0.004    0.426
beta I7    -2.364      0.170   -2.697   -2.031
beta I11    1.420      0.121    1.183    1.658
beta I12   -0.642      0.113   -0.863   -0.422
beta I14    0.663      0.110    0.448    0.879
beta I17   -1.152      0.122   -1.391   -0.913
beta I18    0.565      0.109    0.351    0.778
beta I19   -1.889      0.146   -2.175   -1.602
beta I21   -1.578      0.134   -1.841   -1.315
beta I22   -2.244      0.163   -2.564   -1.925
beta I23    2.103      0.141    1.827    2.379
```

The summary of `rm_sum0` contains two types of parameters: `eta` and `beta`. The `beta` parameter gives the estimated easiness of each item, which is simply the negative of the difficulty parameter `eta`.

As a general note, you will find that many components of the R commands and output will have different names between the different R packages and also compared to the notation used in the theory part of this book. For example, the item parameters are called β in the theory part of the book, `beta` and `eta` (for easiness and difficulty) in this chapter on `eRm`, and again differently in the following chapters on the other R packages. However, we will always explain what is what, and point the reader to the respective sections in the theory part.

We will first look at the first table in the output for the `eta` parameters, representing the item difficulty. In this table, the output starts with the second item. The reason for this is that one constraint is necessary to fix the origin of the latent scale (cf. Section 3.2 and below), so that one item parameter cannot be freely estimated. This is emphasized by omitting the entry for the first item in the table for the `eta` parameters. We also note again that in the FIMS data set in `TAM` only a subset of items is available, so that they are not numbered consecutively.

The second table in the output is for the `beta` parameters, representing the item easiness. Here the values for all items are displayed, including the one for the first item. When you compare the two tables, starting with the second item in the table for the `beta` parameters, you will note that the easiness value is the negative difficulty value, and vice versa. The easiness of the first item, 1.273, thus corresponds to a difficulty of -1.273. We will see below how this value follows from the sum zero constraint, which is used by default for estimating the Rasch model in `eRm`.

We can also access the easiness parameter estimates from an `RM()` result directly by typing

```
> rm_sum0$betapar
```

or

```
> coef(rm_sum0)
   beta I1    beta I2    beta I3    beta I6    beta I7   beta I11
 1.2725717  1.4203230  2.2098398  0.2153106 -2.3639411  1.4203230
  beta I12   beta I14   beta I17   beta I18   beta I19   beta I21
-0.6423858  0.6632561 -1.1517115  0.5646248 -1.8886152 -1.5781132
  beta I22   beta I23
-2.2444186  2.1029365
```

which both give the same output. We can also easily convert the easiness parameters to difficulty parameters by taking their negative, i.e.,

```
> -coef(rm_sum0)
   beta I1    beta I2    beta I3    beta I6    beta I7   beta I11
-1.2725717 -1.4203230 -2.2098398 -0.2153106  2.3639411 -1.4203230
  beta I12   beta I14   beta I17   beta I18   beta I19   beta I21
 0.6423858 -0.6632561  1.1517115 -0.5646248  1.8886152  1.5781132
  beta I22   beta I23
 2.2444186 -2.1029365
```

Through this little detour we can also receive the difficulty parameters of the items, now even including the first item.

For fixing the origin of the latent scale, we can choose between two possible constraints. Our choice is determined by the `sum0` argument to `RM()`. When `sum0 = TRUE`, the sum of all item parameters is constrained to zero. This is the default value for the `sum0` argument, i.e., the value that is used if we do not explicitly specify this argument.

When we first used the `RM()` command above, we did not specify `sum0`, so that by default `sum0 = TRUE` was employed. In this case the value of the first item parameter must be whatever value is missing so that all item parameters together sum up to zero.

If you are curious, you can verify this by typing

```
> sum(rm_sum0$betapar)
[1] -1.110223e-16
```

This value[2], $-1.110223 \cdot 10^{-16}$ in what is called the scientific notation in R, is essentially zero (the tiny deviation being due to rounding errors). If you

[2]Depending on your settings and operating system, the precision with which this result is displayed can vary.

increase the penalty for the scientific notation `scipen` to a large number, for example

```
> options(scipen = 20)
```

and compute the sum again with

```
> sum(rm_sum0$betapar)
[1] -0.00000000000000001110223
```

you can see that it is indeed very close to zero.

When we choose the other constraint option by means of `sum0 = FALSE`, the item parameter of the first item is set to zero. This is demonstrated below.

```
> rm_first0 <- RM(responses, sum0 = FALSE)
> summary(rm_first0)

Results of RM estimation:

Call:  RM(X = responses, sum0 = FALSE)

Conditional log-likelihood: -1886.529
Number of iterations: 16
Number of parameters: 13

Item (Category) Difficulty Parameters (eta): with 0.95 CI:
    Estimate Std. Error lower CI upper CI
I2    -0.148      0.172   -0.485    0.189
I3    -0.937      0.192   -1.313   -0.562
I6     1.057      0.163    0.738    1.377
I7     3.637      0.218    3.209    4.064
I11   -0.148      0.172   -0.485    0.189
I12    1.915      0.168    1.585    2.245
I14    0.609      0.164    0.288    0.931
I17    2.424      0.176    2.079    2.770
I18    0.708      0.164    0.387    1.029
I19    3.161      0.197    2.775    3.547
I21    2.851      0.187    2.485    3.216
I22    3.517      0.212    3.101    3.933
I23   -0.830      0.188   -1.199   -0.462

Item Easiness Parameters (beta) with 0.95 CI:
            Estimate Std. Error lower CI upper CI
beta I1       0.000      0.000    0.000    0.000
beta I2       0.148      0.172   -0.189    0.485
```

```
beta I3     0.937       0.192    0.562     1.313
beta I6    -1.057       0.163   -1.377    -0.738
beta I7    -3.637       0.218   -4.064    -3.209
beta I11    0.148       0.172   -0.189     0.485
beta I12   -1.915       0.168   -2.245    -1.585
beta I14   -0.609       0.164   -0.931    -0.288
beta I17   -2.424       0.176   -2.770    -2.079
beta I18   -0.708       0.164   -1.029    -0.387
beta I19   -3.161       0.197   -3.547    -2.775
beta I21   -2.851       0.187   -3.216    -2.485
beta I22   -3.517       0.212   -3.933    -3.101
beta I23    0.830       0.188    0.462     1.199
```

We can see in the second table with the `beta` parameters that now the first easiness parameter has a value of zero as intended. We can conclude that the first difficulty or `eta` parameter (again not displayed in this output) has a value of zero as well.

It is straightforward to transform between the two parameterizations. Suppose we have used the default `sum0 = TRUE`, where the sum of all item parameters is zero. We can transform the item parameters to the second parameterization, where the first item parameter is zero, by subtracting the value of the first item parameter from all item parameters, i.e.,

```
> comp_betapar_first0 <- rm_sum0$betapar - rm_sum0$betapar[1]
```

By collecting the two different estimates and the transformed values into a `data.frame`, we can verify that the values for the item parameters estimated with the first constraint and transformed by subtracting the first item parameter and those directly estimated with the second constraint are nearly identical to each other.[3]

```
> data.frame(est_sum0 = rm_sum0$betapar,
+            transf = comp_betapar_first0,
+            est_first0 = rm_first0$betapar)
          est_sum0    transf est_first0
beta I1  1.2725717 0.0000000  0.0000000
beta I2  1.4203230 0.1477512  0.1477300
beta I3  2.2098398 0.9372681  0.9372521
```

[3]In `eRm`, the item parameters are estimated by means of an iterative algorithm that generally stops as soon as a convergence criterion is met or a maximum number of computation steps is reached. We could increase the estimation accuracy a little bit by making the criterion stricter. This would further reduce the small differences between the two item parameter estimates, which are caused by practically negligible numerical inaccuracies. However, the necessary commands are very technical and offer no practical advantages, therefore we do not present them here.

```
beta I6    0.2153106 -1.0572611 -1.0572790
beta I7   -2.3639411 -3.6365129 -3.6365367
beta I11   1.4203230  0.1477512  0.1477300
beta I12 -0.6423858 -1.9149575 -1.9149741
beta I14   0.6632561 -0.6093156 -0.6093268
beta I17 -1.1517115 -2.4242832 -2.4243012
beta I18   0.5646248 -0.7079469 -0.7079591
beta I19 -1.8886152 -3.1611869 -3.1612110
beta I21 -1.5781132 -2.8506849 -2.8506972
beta I22 -2.2444186 -3.5169903 -3.5170051
beta I23   2.1029365  0.8303648  0.8303438
```

Alternatively, if we have set `sum0 = FALSE`, we can force the item parameters to sum to zero by subtracting the mean of the `beta` parameters from each value, i.e.,

```
> comp_betapar_sum0 <- rm_first0$betapar - mean(rm_first0$betapar)
```

We can verify this in the same way we did for the previous transformation by collecting the estimates and transformed values into a `data.frame`.

```
> data.frame(est_sum0 = rm_sum0$betapar,
+            transf = comp_betapar_sum0,
+            est_first0 = rm_first0$betapar)
           est_sum0     transf est_first0
beta I1    1.2725717  1.2725882  0.0000000
beta I2    1.4203230  1.4203181  0.1477300
beta I3    2.2098398  2.2098403  0.9372521
beta I6    0.2153106  0.2153092 -1.0572790
beta I7   -2.3639411 -2.3639485 -3.6365367
beta I11   1.4203230  1.4203181  0.1477300
beta I12 -0.6423858 -0.6423860 -1.9149741
beta I14   0.6632561  0.6632614 -0.6093268
beta I17 -1.1517115 -1.1517130 -2.4243012
beta I18   0.5646248  0.5646290 -0.7079591
beta I19 -1.8886152 -1.8886229 -3.1612110
beta I21 -1.5781132 -1.5781090 -2.8506972
beta I22 -2.2444186 -2.2444169 -3.5170051
beta I23   2.1029365  2.1029320  0.8303438
```

We can observe the relationship between the number of times an item was correctly answered (the column sums c_i from Chapter 2, termed item scores in the following to avoid confusion with the R command `colSums`) and the estimated item parameters by creating a **data.frame** containing both the item scores and their associated item parameter estimates.

```
> tab <- data.frame(item_score = colSums(responses),
+                   easiness = coef(rm_sum0),
+                   difficulty = -coef(rm_sum0))
```

We can highlight the relationship between item score and difficulty by ordering the rows of this data frame, which we have stored under the name **tab**, by the item score

```
> tab[order(tab$item_score), ]
    item_score   easiness difficulty
I7          40 -2.3639411  2.3639411
I22         44 -2.2444186  2.2444186
I19         58 -1.8886152  1.8886152
I21         73 -1.5781132  1.5781132
I17         98 -1.1517115  1.1517115
I12        134 -0.6423858  0.6423858
I6         204  0.2153106 -0.2153106
I18        233  0.5646248 -0.5646248
I14        241  0.6632561 -0.6632561
I1         287  1.2725717 -1.2725717
I2         297  1.4203230 -1.4203230
I11        297  1.4203230 -1.4203230
I23        336  2.1029365 -2.1029365
I3         341  2.2098398 -2.2098398
```

Looking at this table, we see that items that were correctly answered by only few test takers have low easiness and high difficulty parameters. Intuitively, this relationship makes sense. Under the Rasch model the probability of correctly answering an item decreases with difficulty. Thus, the items with the fewest correct responses should be inferred to be the most difficult. By contrast, those items that were correctly answered by many test takers have high easiness and low difficulty parameters.

6.2 Test Evaluation

We will now go through some of the means for test evaluation that we have introduced in Chapter 4.

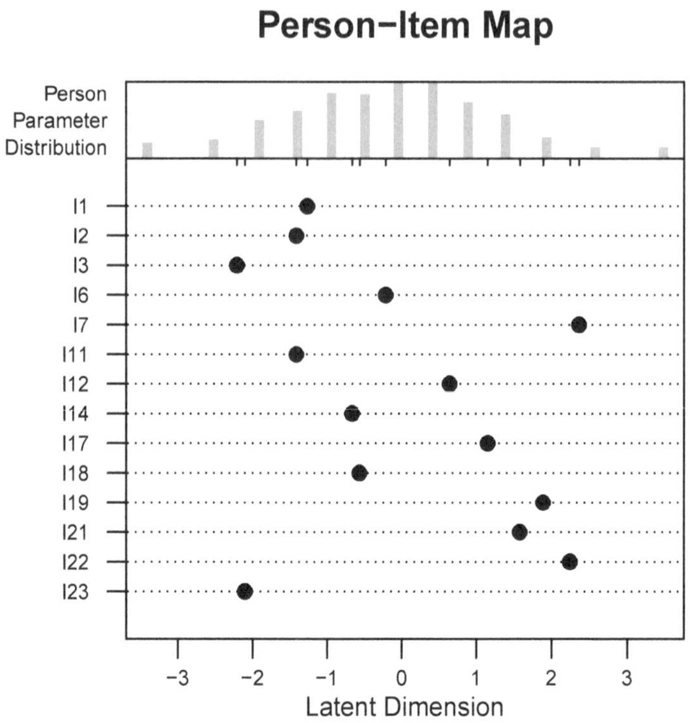

FIGURE 6.1: Person item map from the FIMS estimates.

6.2.1 Person Item Map

Let us first check whether the locations of persons and items on the latent continuum are well aligned. To create the person item map (cf. Section 4.1.1), type

```
> plotPImap(rm_sum0)
```

(In case you get the warning that the figure margins are too large, you can manually increase the size of the plot window in RStudio or use the command `dev.new()` before creating the plot.)

Figure 6.1 shows a person item map for our test responses. The upper portion of the person item map displays a histogram of ability parameter estimates, and the lower portion displays the difficulty estimates for each test item. For each item, the difficulty estimate is denoted by the position of the point on the dashed line corresponding to that item. For example, the difficulty estimated for item 1 corresponds to the location of the point on the uppermost dashed line.

The person item map offers a visual sanity check for the estimates of our IRT model. Ability estimates are most accurate when they fall in the middle of the distribution of item parameters, and vice versa. Thus, the ability histogram and difficulty estimates should ideally be centered on the same point and show a large overlap. For our test, this appears to be the case.

6.2.2 Empirical ICCs

Next we will visually inspect the empirical and expected ICCs (cf. Section 4.1.2). We can display them for all items at the same time using the command

```
> plotICC(rm_sum0, item.subset = "all",
+         empICC = list("raw"), empCI = list())
```

where `empICC` adds the points for the empirical ICCs (i.e., estimated probabilities of solving each item for different values of the latent trait) and `empCI` adds a confidence intervals for each point estimate (which are larger in non-central regions of the latent dimension, where there are few persons, cf. person item map and Section 4.1.1).

To save space, we will only display the plots for four exemplary items in Figure 6.2 by means of specifying their names in the `item.subset` argument

```
> plotICC(rm_sum0, item.subset = c("I12", "I14",
+         "I19", "I21"), empICC = list("raw"),
+         empCI = list())
```

The empirical ICCs are indicated by the individual points, the expected ICC under the Rasch model is indicated by the smooth line. From the items displayed in Figure 6.2, for items 12 and 14 we note that in general the shape of the empirical ICC is very well aligned with the expected ICC, but for item 12 the empirical ICC shows values above zero even for the smallest abilities to the left of the latent dimension. This could indicate guessing and will be further investigated in Chapter 7. For item 19 the empirical ICC looks steeper than the expected ICC under the Rasch model. It shows a much more pronounced jump between the first roughly half of the points and the remaining points. For item 21, by contrast, the empirical ICC is much flatter than expected. We will compare our visual impression with the item fit statistics for these items below.

6.2.3 Andersen's Likelihood Ratio Test and Graphical Test

The next means for graphically assessing the Rasch model, which we have presented in Section 4.1.3, is the graphical test. In `eRm`, the function for the graphical test "recycles" the result from Andersen's likelihood ratio test, which

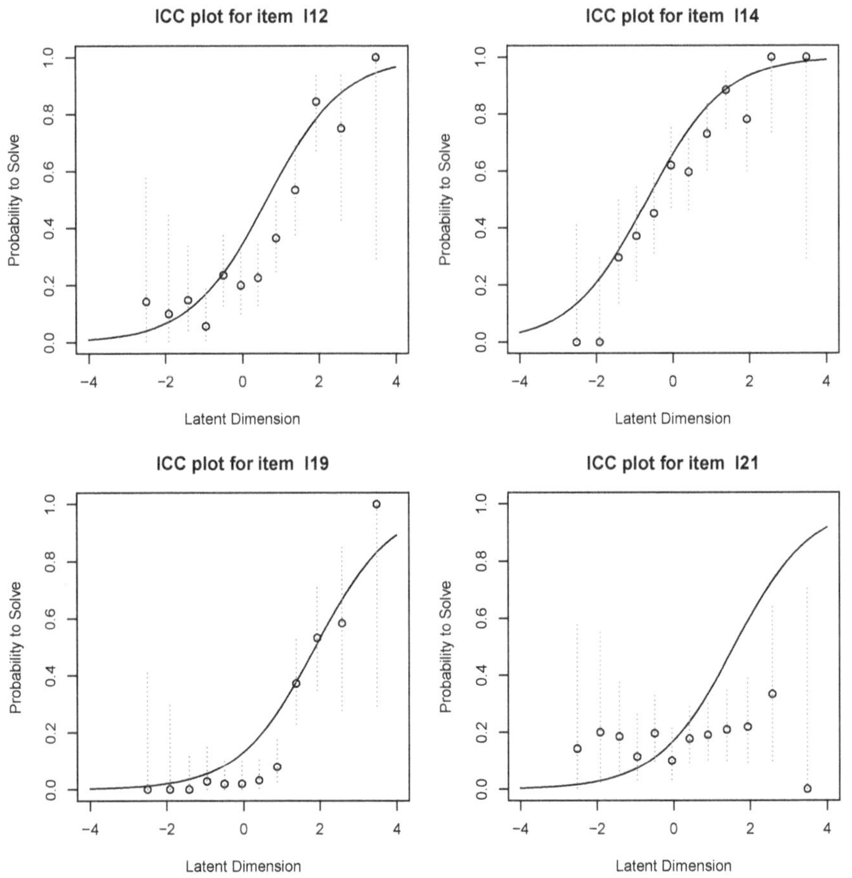

FIGURE 6.2: Empirical and expected ICCs for exemplary items.

we presented in Section 4.2.1. This is why we will show how to conduct this test first.

Andersen's likelihood ratio test is provided in `eRm` through the `LRtest()` function. Initially, we will test whether item parameters differ for test takers with below or above average scores. This test can be performed in R by calling `LRtest()` on the result from `RM()` with `splitcr = "mean"`, i.e.,

```
> lrt_mean_split <- LRtest(rm_sum0, splitcr = "mean")
> lrt_mean_split

Andersen LR-test:
LR-value: 79.71
```

```
Chi-square df: 13
p-value:  0
```

The output from this test shows a significant violation of the Rasch model at the $\alpha = 0.05$ level.[4]

We can also provide the `splitcr` argument with a variable dividing test takers into groups. Here we will test whether the item parameters differ across gender. We can do this by passing a vector containing group memberships as the `splitcr` argument:

```
> lrt_gender <- LRtest(rm_sum0, splitcr = gender)
> lrt_gender

Andersen LR-test:
LR-value: 32.973
Chi-square df: 13
p-value:  0.002
```

Note that here `gender` is not written in quotes, while in the previous command the split criterion `"mean"` was written in quotes. The reason for this is that `"mean"` is the keyword for choosing a split criterion that is already implemented as a general rule in the `LRtest()` function, while `gender` is the vector of group memberships we have just created.

As with the previous test, the LRT for gender also indicates a significant violation of the Rasch model at the $\alpha = 0.05$ level.

We can now plot the difficulty estimates of each group against each other using the `plotGOF()` function to create the graphical test discussed in Section 4.1.3. The `plotGOF()` function takes the result of `LRtest()` and plots the item parameter estimates for the two groups. To help with visual assessment, `plotGOF()` can optionally label the items and add confidence ellipses. The `tlab` and `pos` arguments control the way labelling is performed and the `conf` argument controls how the confidence ellipses are displayed. For example, we can plot the estimates for test takers scoring below the mean score against those scoring above the mean score, along with numeric labels and 95% confidence ellipses, as follows. The resulting plots are shown in Figure 6.3 and Figure 6.4.

We can create the plot for the graphical test based on the mean split (Figure 6.3) with

```
> plotGOF(lrt_mean_split,
+         tlab = "item", pos = 1,
+         main = "Difficulty by Score (with Item Names)",
+         conf = list(gamma = 0.95, col = 1))
```

[4]The p-value is not truly zero. **eRm** simply rounds the output to the third decimal place.

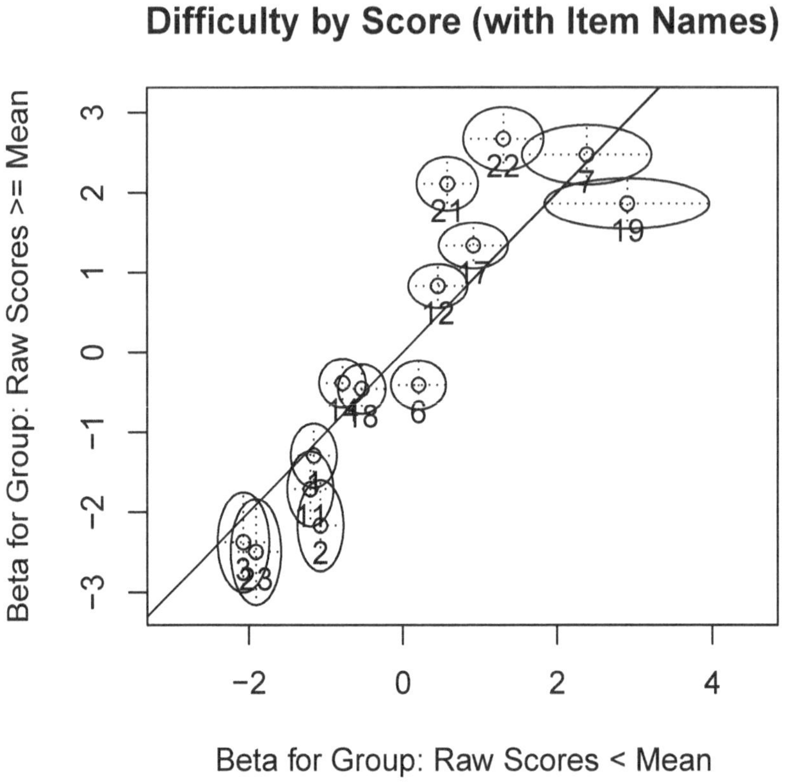

FIGURE 6.3: Difficulty parameter estimates for test takers scoring below and above the mean.

and the plot for the graphical test based on gender (Figure 6.4) with

```
> plotGOF(lrt_gender,
+         tlab = "item", pos = 1,
+         main = "Difficulty by Gender (with Item Names)",
+         conf = list(gamma = 0.95, col = 1))
```

Each small circle in Figure 6.3 and Figure 6.4 shows the difficulty estimates for a single item. A circle's x-coordinate in Figure 6.3 gives its difficulty estimate for test takers scoring below the mean and its y-coordinate gives its difficulty estimate for test takers scoring above the mean. The line $y = x$ is provided as a reference, since points falling on this line would have the same estimate in both groups. The distance between any point and the reference line $y = x$ indicates how much estimates differ between the two groups. It also indicates the direction of this difference. Items below the line are more

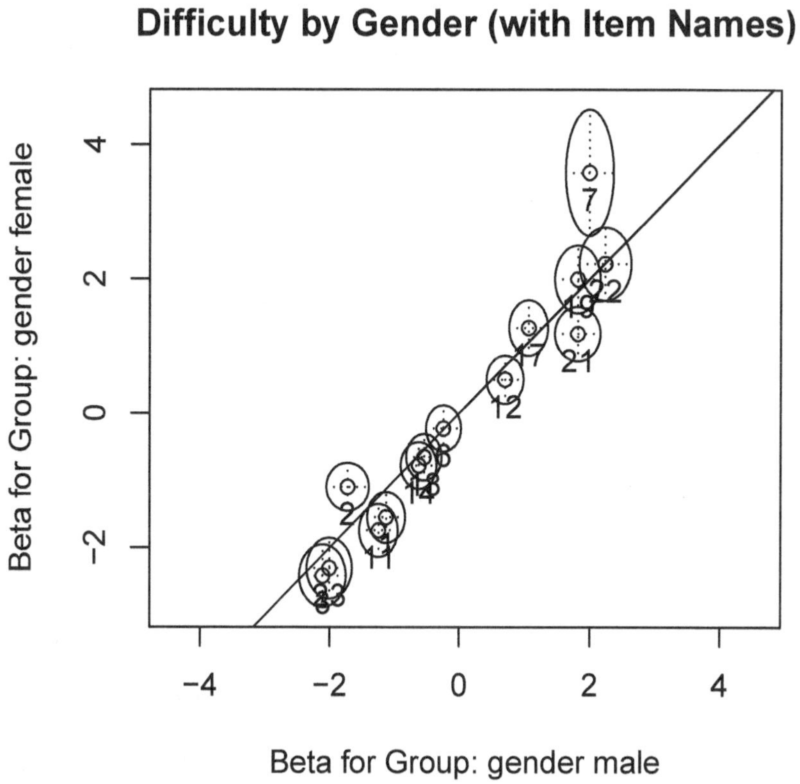

FIGURE 6.4: Difficulty parameter estimates for female and male test takers.

difficult for test takers scoring below the mean, while items above the line are more difficult for test takers scoring above the mean.

However, as we have seen in Chapter 4, this reference line is useful for judging whether individual items have DIF only when DIF is balanced and cancels out across all items. Whether this is the case here is hard to tell at first glance. For now, we will stick to this reference line and use the confidence ellipses to judge the magnitude of deviations from it. The horizontal and vertical axes show confidence intervals for the estimates for each group of test takers. The width of each confidence interval is determined by the `gamma` element of the list provided to `conf`. The default setting `gamma = .95` results in 95% confidence intervals for each of the ellipse's axes, drawn in black here with `col = 1`. When a confidence ellipse does not cross the reference line the respective item is diagnosed as showing significant DIF.

Figure 6.3 indicates that items 2, 6, 21, and 22 differ significantly between people scoring above and below the mean, since their confidence ellipses do not cross the reference line. Items 21 and 22 are more difficult for people scoring at or above the mean, while items 2 and 6 are more difficult for people scoring below the mean. Such a model violations can occur when the observed ICCs differ from the ICCs expected under the Rasch model for test takers with low and high abilities. This may happen, for instance, if guessing is present, or if the slope is steeper or less steep than predicted by the Rasch model.

Additionally, Figure 6.4 indicates that items 2, 7, and 21 differ between female and male test takers. Items 2 and 7 are more difficult for female test takers, while item 21 is more difficult for male test takers.

Note that in the R command for the graphical test above, we have indicated by `tlab = "item"` that the item names should be used as their labels. An alternative would be to use the item numbers by means of `tlab = "number"`. The resulting plot is shown in Figure 6.5.

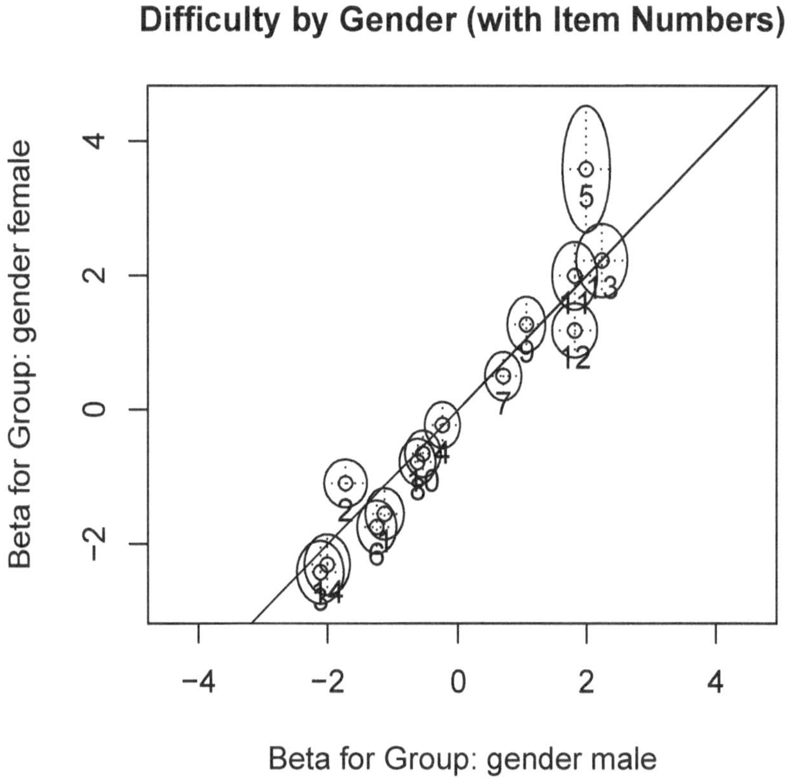

FIGURE 6.5: Difficulty parameter estimates for female and male test takers. Items indicated by item numbers instead of item names.

As we have already pointed out, in this data set the items are not numbered consecutively. For example, we can see in Figure 6.5 that the fifth item (item number) lies above the diagonal. This corresponds to item 7 (item name) in Figure 6.4. We are not pointing this out to confuse the reader, but to emphasize how important it is to be aware of the difference between item names and item numbers for correctly interpreting the results in case your own data contain items that are not numbered consecutively.

6.2.4 Wald Test

As discussed in Section 4.2.3, we can also use the item-wise Wald test to test for group differences in individual difficulty parameters. The `eRm` package provides this functionality through the `Waldtest()` function. For example, we can test for differences between test takers scoring above and below the mean as follows.

```
> Waldtest(rm_sum0, splitcr = "mean")

Wald test on item level (z-values):

          z-statistic p-value
beta I1        -0.514   0.607
beta I2        -3.328   0.001
beta I3        -0.838   0.402
beta I6        -2.555   0.011
beta I7         0.210   0.834
beta I11       -1.773   0.076
beta I12        1.562   0.118
beta I14        1.821   0.069
beta I17        1.550   0.121
beta I18        0.333   0.739
beta I19       -1.827   0.068
beta I21        5.768   0.000
beta I22        4.106   0.000
beta I23       -1.560   0.119
```

These tests again indicate that items 2, 6, 21, and 22 differ significantly across test takers scoring above and below the mean, consistent with Figure 6.3. These items will be further investigated in Chapter 7.

We can similarly test for differences between females and males by typing

```
> Waldtest(rm_sum0, splitcr = gender)

Wald test on item level (z-values):

          z-statistic p-value
beta I1        -1.727   0.084
beta I2         2.543   0.011
```

```
beta I3          -1.020   0.308
beta I6           0.067   0.946
beta I7           3.089   0.002
beta I11         -1.978   0.048
beta I12         -0.861   0.389
beta I14         -0.673   0.501
beta I17          0.815   0.415
beta I18         -0.493   0.622
beta I19          0.583   0.560
beta I21         -2.305   0.021
beta I22         -0.030   0.976
beta I23         -1.019   0.308
```

The results here also largely agree with Figure 6.4. Consistent with the graphical test, the Wald test indicates that items 2, 7, and 21 differ between the groups. Unlike the graphical test, the Wald test also indicates that item 11 differs between the groups. The reason for this apparent disagreement is that for the Wald test, we can clearly tell that the p-value 0.048 is below 5%. For the graphical test, on the other hand, it is hard to tell whether the edge of the confidence ellipse for item 11 touches the reference line or not.

6.2.5 Anchoring

Besides the Wald test in `eRm`, which implicitly employs the `sum0 = TRUE` constraint for anchoring (cf. Section 4.2.3 and Section 4.2.4), we would like to shortly point out additional anchoring options provided by the `psychotools` package (Zeileis, Strobl, Wickelmaier, Komboz, Kopf, Schneider, & Debelak, 2021). For the sake of brevity, we will only repeat the DIF analysis for gender.

After installing and loading the `psychotools` package with

```
> install.packages("psychotools")
> library("psychotools")
```

we prepare the item responses in matrix form as required by `psychotools`

```
> resp <- as.matrix(responses)
```

and conduct a Wald test for gender based on different anchor selection approaches. We start with using the `constant` four mean p-value threshold (`MPT`) approach (Kopf, Zeileis, & Strobl, 2015a), which selects anchor items based on a significance threshold

```
> anchortest(resp ~ gender,
+            class = "constant",
+            select = "MPT")
```

```
Anchor items:
respI23, respI22, respI3, respI18

Final DIF tests:

 Simultaneous Tests for General Linear Hypotheses

Linear Hypotheses:
             Estimate Std. Error z value Pr(>|z|)
respI1 == 0   0.24221    0.29004   0.835  0.40368
respI2 == 0  -0.81096    0.29229  -2.775  0.00553 **
respI3 == 0   0.12532    0.26670   0.470  0.63844
respI6 == 0  -0.19638    0.27092  -0.725  0.46853
respI7 == 0  -1.77446    0.57169  -3.104  0.00191 **
respI11 == 0  0.31848    0.29708   1.072  0.28371
respI12 == 0  0.01949    0.28209   0.069  0.94491
respI14 == 0 -0.02886    0.27344  -0.106  0.91596
respI17 == 0 -0.39256    0.30802  -1.274  0.20251
respI18 == 0 -0.07045    0.22034  -0.320  0.74916
respI19 == 0 -0.36408    0.36215  -1.005  0.31473
respI21 == 0  0.44768    0.32158   1.392  0.16388
respI22 == 0 -0.17097    0.30188  -0.566  0.57115
---
Signif. codes:  0 '***' 0.001 '**' 0.01 '*' 0.05 '.' 0.1 ' ' 1
(Univariate p values reported)
```

and the iterative **forward** mean test statistic threshold (MTT) method (Kopf et al., 2015b), which iteratively adds items to the anchor.

```
> anchortest(resp ~ gender,
+            class = "forward",
+            select = "MTT")
Anchor items:
respI23, respI12, respI14, respI18, respI1, respI6,
respI11, respI17, respI19

Final DIF tests:

 Simultaneous Tests for General Linear Hypotheses

Linear Hypotheses:
             Estimate Std. Error z value Pr(>|z|)
respI1 == 0   0.28177    0.23616   1.193  0.23283
respI2 == 0  -0.77140    0.26596  -2.900  0.00373 **
respI3 == 0   0.16488    0.32294   0.511  0.60967
```

```
respI6 == 0  -0.15682   0.21567  -0.727  0.46715
respI7 == 0  -1.73490   0.55736  -3.113  0.00185 **
respI11 == 0  0.35804   0.24322   1.472  0.14099
respI12 == 0  0.05905   0.22616   0.261  0.79401
respI14 == 0  0.01070   0.21884   0.049  0.96099
respI17 == 0 -0.35299   0.25138  -1.404  0.16026
respI18 == 0 -0.03089   0.21750  -0.142  0.88706
respI19 == 0 -0.32452   0.30298  -1.071  0.28412
respI21 == 0  0.48724   0.29524   1.650  0.09887 .
respI22 == 0 -0.13141   0.37175  -0.353  0.72372
---
Signif. codes:
0 '***' 0.001 '**' 0.01 '*' 0.05 '.' 0.1 ' ' 1
(Univariate p values reported)
```

These two approaches performed particularly well in the simulation studies of Kopf et al. (2015a) and Kopf et al. (2015b). Another alternative available in **psychotools** is the anchor point selection approach of Strobl, Kopf, Kohler, von Oertzen, and Zeileis (2021), which is based on optimizing the `Gini` index.

```
> anchortest(resp ~ gender,
+            select = "Gini")
Anchor items:
respI23

Final DIF tests:

 Simultaneous Tests for General Linear Hypotheses

Linear Hypotheses:
              Estimate Std. Error z value Pr(>|z|)
respI1 == 0   0.126100   0.387369   0.326  0.74478
respI2 == 0  -0.927061   0.389044  -2.383  0.01718 *
respI3 == 0   0.009212   0.427183   0.022  0.98280
respI6 == 0  -0.312486   0.376562  -0.830  0.40663
respI7 == 0  -1.890565   0.632101  -2.991  0.00278 **
respI11 == 0  0.202375   0.392278   0.516  0.60593
respI12 == 0 -0.096612   0.386789  -0.250  0.80276
respI14 == 0 -0.144961   0.376949  -0.385  0.70056
respI17 == 0 -0.508660   0.407380  -1.249  0.21181
respI18 == 0 -0.186557   0.376409  -0.496  0.62016
respI19 == 0 -0.480188   0.450796  -1.065  0.28679
respI21 == 0  0.331579   0.418092   0.793  0.42773
respI22 == 0 -0.287076   0.476210  -0.603  0.54662
---
```

```
Signif. codes:
0 '***' 0.001 '**' 0.01 '*' 0.05 '.' 0.1 ' ' 1
(Univariate p values reported)
```

The R outputs from the **anchortest()** function displayed above list the anchor item(s) selected by the respective anchor selection approach, as well as the results for the item-wise Wald test from Section 4.2.3 based on these anchor items. All three approaches lead to results where only items 2 and 7 show DIF for gender, whereas the graphical test and the Wald test in **eRm** also identified item 21 and item 11 (borderline) as having DIF.

When we look again at the graphical test in Figure 6.4, we see that items 2 and 7 show DIF in the same direction (above the diagonal), while items 21 and 11 point in the other direction (below the diagonal), and to a lesser degree.

Taken together, these findings indicate that unbalanced DIF might be present, and that the diagonal used in Figure 6.4 is not ideal for assessing the items. To illustrate this, we manually draw an alternative reference line through the location of item 23, which was selected as the (first) anchor item by the three approaches from **psychotools** presented above[5], by means of the **abline** command.

```
> plotGOF(lrt_gender,
+         tlab = "item", pos = 1,
+         main = "Difficulty by Gender (with Item Names)",
+         conf = list(gamma = 0.95, col = 1))
> abline(-0.3, 1, lty=2)
```

As can be seen in the resulting Figure 6.6, based on the alternative reference line we no longer find DIF in items 11 and 21, but items 2 and 7 show DIF even more clearly.

For this data example, the same conclusion is reached in **eRm** when employing the **stepwiseIt()** function, that conducts several Wald tests and in each step excludes the single item with the largest test statistic.

[5]When you look back at the R outputs from the **anchortest()** function displayed above, you will also notice that no Wald test result is listed for this item. Mathematically speaking, this is a consequence of the fact that we need one restriction for estimating the item parameters, so that for a test with I items only $I-1$ standard errors can be estimated. The **anchortest()** function handles this by transforming the covariance matrices accordingly (for details see Kopf et al., 2015b).
Not testing this item for DIF also makes sense intuitively, since the item that is first selected into the anchor is the one that is least likely to have DIF. Therefore, all items that are not flagged by **anchortest()** – including the item for which no test result is listed – can be interpreted as showing no significant DIF.

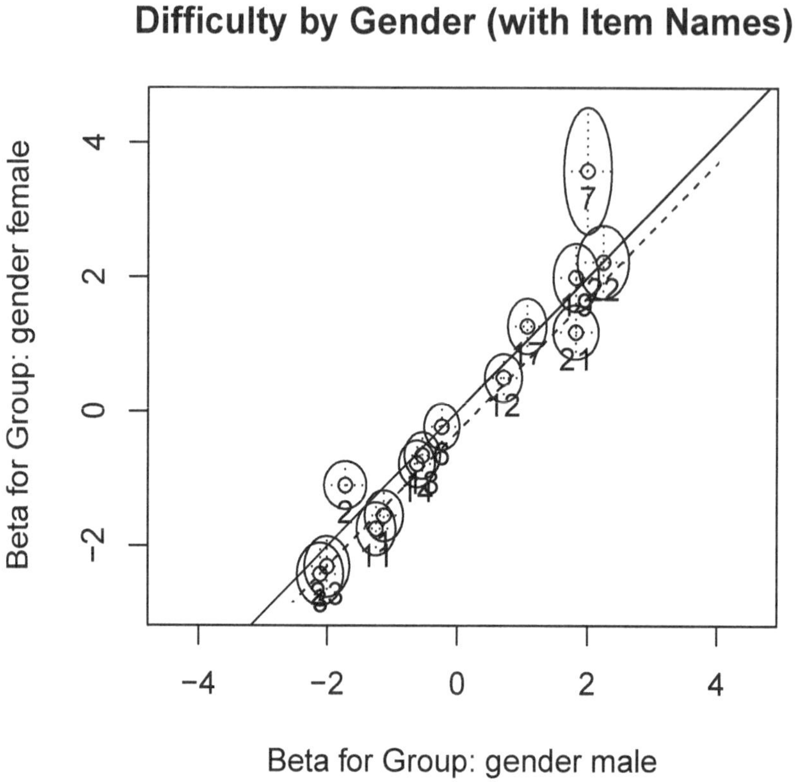

FIGURE 6.6: Difficulty parameter estimates for female and male test takers with alternative reference line (dashed line).

```
> stepwiseIt(rm_sum0, criterion = list("Waldtest", gender))
Eliminated item - Step 1: I7
Eliminated item - Step 2: I2

Results for stepwise item elimination:
Number of steps: 2
Criterion: Waldtest

             z-statistic p-value
Step 1: I7         3.089   0.002
Step 2: I2         3.059   0.002
```

With this approach, after the two items with the largest DIF, items 7 and 2, have been excluded, the remaining items don't show any more significant DIF. To understand this result, you can imagine that the reference line in Figure 6.6 would be the solid diagonal line in the first step, but after removing item 7 would move towards the dashed line in the second step, and after further removing item 2 would be situated at or very close to the dashed line in the third step, so that the remaining items show no more significant DIF.

To conclude with this topic: While we have seen that the naive graphical and Wald tests based on the sum zero constraint can be misled by unbalanced DIF, modern anchoring approaches and the stepwise item elimination approach can provide a clearer picture.

6.2.6 Removing Problematic Items

If this analysis were part of a real test construction, those items showing DIF (or other anomalies in our later analyses) would need to be closely investigated by content experts to decide whether they should be modified or removed from the test, as discussed in Section 4.2.6. Here we will not actually exclude items because we want to retain the full data set for all parts of the illustration and for comparability between the different R packages. However, if you wanted to remove items (i.e., columns) from your data set, this could be done with the following commands.

For example, for removing the second and fifth column from the data set, we could use the command

```
> responses_removeDIFitems <- responses[, -c(2, 5)]
```

In this particular data set, however, the items are not numbered consecutively, so that the column numbers and the numbers in the items' names are not identical. While the second column does contain item 2, item 7 is stored in the fifth column, as we can see, for example, from

```
> colnames(responses)
 [1] "I1"  "I2"  "I3"  "I6"  "I7"  "I11" "I12" "I14" "I17" "I18"
[11] "I19" "I21" "I22" "I23"
```

Removing the items by negative indexing of the column numbers like above can be error-prone in this situation. Note also that if you intend to remove additional items in another command, the column numbers of the remaining items will have changed, even if they agreed originally. An alternative approach that may cause less of a headache is to address the items by means of their names, for example using the following command

```
> responses_removeDIF <- responses[, -which(colnames(responses)
+                                     %in% c("I2", "I7"))]
```

After removing items, the entire process would have to start again from scratch by re-fitting the Rasch model and investigating the remaining items. Here we will instead proceed with the original model including all items to ensure comparability between all examples and chapters.

6.2.7 Martin-Löf Test

Next we will apply the Martin-Löf test to check whether various subgroups of items measure the same latent ability (cf. Section 4.2.2). To check whether the easy and difficult items of the test lead to the same ability parameter estimates, we can use the following code:

```
> mloef_median <- MLoef(rm_sum0, splitcr = "median")
> mloef_median

Martin-Loef-Test (split criterion: median)
LR-value: 67.083
Chi-square df: 48
p-value: 0.036
```

We obtain a p-value below 0.05. This indicates that the person parameter estimates obtained from the easy and difficult items differ significantly, i.e., a violation of the Rasch model.

In this analysis, the items were split with regard to the **median** of the item scores, i.e., the number of correct responses per item. Alternatively, we could also explicitly define item groups which should be compared in the Martin-Löf test. This can be done by defining a vector whose length equals the number of items. Each entry in this vector indicates the group to which it belongs. We don't have enough content knowledge to suspect particular groups of items to measure different latent traits, so we will illustrate the R commands by arbitrarily using the first seven and the last seven items in our data set, which includes 14 items in total, as the two item groups.

```
> split <- c(rep(0, 7), rep(1, 7))
> mloef_alter <- MLoef(rm_sum0, splitcr = split)
> mloef_alter

Martin-Loef-Test (split criterion: user-defined)
LR-value: 70.568
Chi-square df: 48
p-value: 0.019
```

Despite our arbitrary choice of item groups, we obtain a p-value below 0.05, which indicates a violation of the Rasch model. With more content knowledge we could hypothesize which properties of the two item groups might be responsible for this finding. But note that – as always with statistical tests – this could also be a false-positive finding.

6.2.8 Item and Person Fit

Finally, we can obtain fit statistics for individual items and persons. As we have explained in Section 4.3.3, infit and outfit statistics are commonly used for checking the fit of individual items or persons in the Rasch model. As was also mentioned in this section, the calculation of Rasch residuals is based on the estimates of the item and person parameters. The details on the estimation of the person parameters will be explained below. First, we calculate the infit and outfit statistics for all items in our data set using the following command.

```
> itemfit(person.parameter(rm_sum0))

Itemfit Statistics:
      Chisq  df p-value Outfit MSQ Infit MSQ Outfit t Infit t
I1  325.410 397   0.996      0.818     0.904   -1.750  -1.666
I2  273.788 397   1.000      0.688     0.809   -2.928  -3.271
I3  289.188 397   1.000      0.727     0.869   -1.574  -1.505
I6  333.574 397   0.991      0.838     0.860   -2.317  -3.257
I7  272.900 397   1.000      0.686     0.838   -1.328  -1.423
I11 332.540 397   0.992      0.836     0.816   -1.432  -3.143
I12 458.651 397   0.018      1.152     0.972    1.538  -0.538
I14 395.777 397   0.508      0.994     1.023   -0.043   0.492
I17 524.132 397   0.000      1.317     0.936    2.290  -1.031
I18 432.879 397   0.104      1.088     1.019    1.121   0.424
I19 226.655 397   1.000      0.569     0.750   -2.579  -2.999
I21 905.981 397   0.000      2.276     1.246    5.846   2.986
I22 727.958 397   0.000      1.829     0.985    2.918  -0.105
I23 275.909 397   1.000      0.693     0.852   -1.918  -1.812
    Discrim
I1    0.418
I2    0.520
I3    0.371
I6    0.505
I7    0.279
I11   0.473
I12   0.321
I14   0.320
I17   0.280
I18   0.314
I19   0.453
I21  -0.108
I22   0.059
I23   0.412
```

The resulting table first contains the test statistics of an approximate χ^2 goodness-of-fit test, its degrees of freedom, and the resulting p-values. The reasoning behind these tests was described in Section 4.3.1, and the calculation of these statistics was described in Section 4.3.3. As was stated in this section, the resulting test statistic can be roughly approximated by a χ^2 distribution if the Rasch model holds, which leads to the presented p-values.[6]

The following columns present infit and outfit MSQ and t statistics. We remind the reader that for the infit and outfit MSQ statistics, values close to 1 indicate a good model fit, while for the infit and outfit t statistics, values close to 0 indicate a good fit. As explained in Section 4.3.3, higher values indicate that responses are more random than expected under the Rasch model, which indicates underfit. Lower values indicate that responses are less random than expected, which indicates overfit.

Following one of the guidelines reported in Section 4.3.3, we will further inspect those items whose infit or outfit t values are below -2 or above 2, but would like to emphasize that alternative guidelines exist. We find that for items 2, 6, 11, and 19, at least one t value is below -2. This indicates overfit. For item 19, which we have already visually inspected in Figure 6.2, this is supported by the fact that the empirical ICC has a steeper slope than the expected ICC.

For items 17, 21, and 22, on the other hand, at least one t value for the infit and outfit statistics is above 2, indicating underfit. This again is in line with Figure 6.2, where we found that the empirical ICC for item 21 has a lower slope than the expected ICC. A table with infit and outfit statistics for each person can be obtained with the following code (here we show only the first ten lines to save space):

```
> personfit(person.parameter(rm_sum0))


      Chisq df p-value Outfit MSQ Infit MSQ Outfit t Infit t
1    13.191 13   0.433      0.942     1.080     0.07    0.35
2     7.004 13   0.902      0.500     0.744    -0.44   -0.85
3     7.564 13   0.871      0.540     0.799    -0.37   -0.62
4    13.651 13   0.399      0.975     1.112     0.14    0.44
5     4.024 13   0.991      0.287     0.354    -1.81   -2.20
6    22.049 13   0.055      1.575     1.304     0.96    1.01
7    22.823 13   0.044      1.630     1.623     1.02    1.83
8    12.937 13   0.453      0.924     0.746     0.00   -0.66
9    71.924 13   0.000      5.137     1.376     2.32    1.11
10   22.507 13   0.048      1.608     1.065     1.12    0.30
```

[6] As we have seen in Section 4.3.3, this χ^2 statistic is equal to the numerator of the outfit MSQ statistic. Accordingly, you will find that for each item, its χ^2 value is roughly its outfit MSQ value times the number of persons, excluding respondents with a perfect score and a score of 0. In this data set, this leads to the exclusion of two respondents.

As was the case for the infit and outfit statistics for individual items, persons whose t values are above 2 show response behavior that is more random than expected under the Rasch model (Engelhard, 2013), which could indicate, for instance, guessing or careless response behavior (Bond & Fox, 2007). Following Engelhard (2013), response patterns leading to t values below -2 should also be investigated. They indicate a response behavior that is more deterministic than expected. In this example, persons 5 and 9 would be flagged by these rules of thumb.

6.3 Plots of ICCs, Item and Test Information

We can display the expected ICCs of all test items in a single plot using the `plotjointICC()` function. Here the `cex` argument can be used to regulate the font size for the legend. The resulting plot is shown in Figure 6.7.

```
> plotjointICC(rm_sum0, cex = 0.7)
> segments(-2.5, 0.5, 4, 0.5, lty = 2)
```

We can use this plot to review how difficulty influences the probability that a test taker correctly responds to an item. Recall that an item's difficulty is the ability at which a person has a 50/50 chance of correctly answering the item. We have added a horizontal dashed line at probability 0.5 by means of the `segments` command. The ability at which an ICC crosses this line is its difficulty. This allows us to easily read off the relative difficulties of the items from the graph. Moving from left to right, the first ICC hit by the horizontal line is for the least difficult item (in this case item 3, closely followed by item 23; cf. the item order in the legend and the estimated item parameters in Section 6.1), and the last ICC hit by the horizontal line is for the most difficult item (in this case item 7).

Note that the expected ICCs in Figure 6.7 are parallel by definition. The Rasch model assumes that ICCs are parallel, so it will always produce parallel expected or theoretical ICCs, even when the items have different slopes or guessing rates in reality, as we have seen in Section 4.1.2 and above for the empirical ICCs.

As was explained in Section 3.6, the locations of the ICCs also determine the regions on the latent trait where each item has the highest information. This can be illustrated by means of an item information plot (displayed in Figure 6.8) by typing

```
> plotINFO(rm_sum0, type = "item", legpos = FALSE)
```

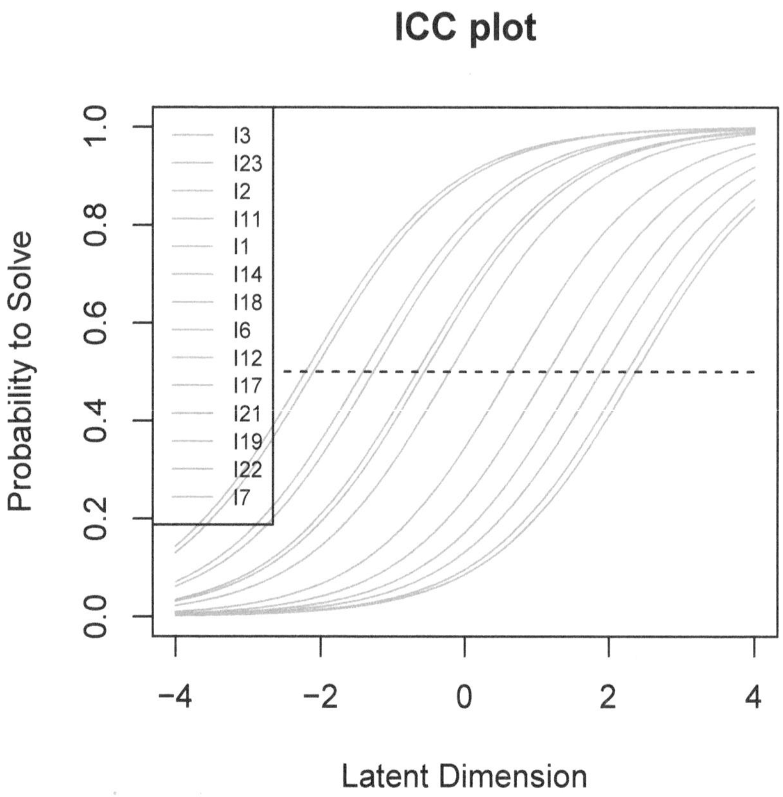

FIGURE 6.7: ICCs for each of the item parameters estimated from the FIMS data.

The information all items together provide for different regions of the latent trait is summarized in the test information plot (displayed in Figure 6.9), which can be generated with

```
> plotINFO(rm_sum0, type = "test")
```

As we have already seen in the person item map, most items are located around zero. As a result, the test information in Figure 6.9 is highest around zero. We also note that the test information curve is slightly skewed and higher to the left of zero than to the right. This is because there are a few more items with their difficulty parameter – and thus their location of highest information – located to the left of zero, as we can also see in the item information plot (Figure 6.8). Since the test information is the sum over the information of the individual items (cf. Section 3.6), the test information is higher in this region.

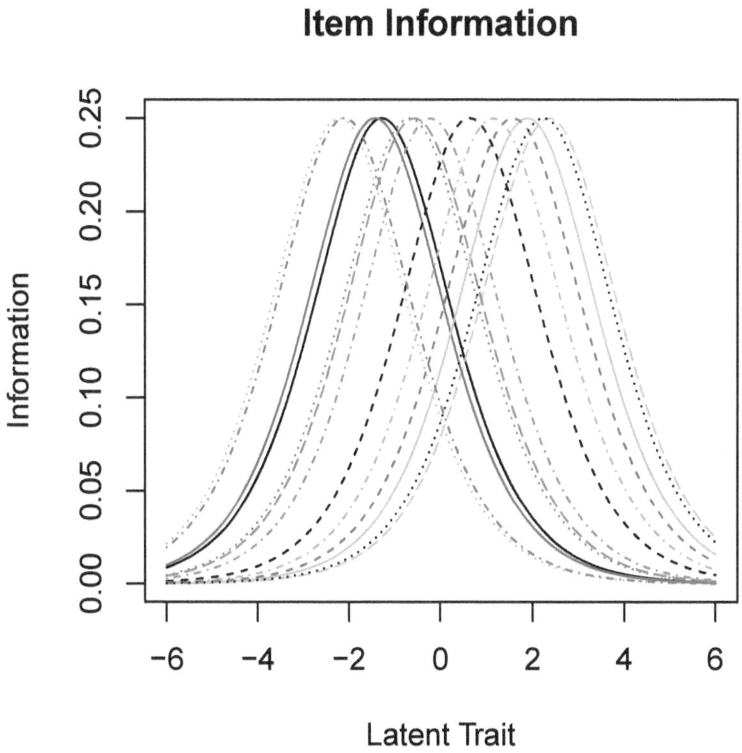

FIGURE 6.8: Item information curves.

6.4 Person Parameter Estimation

To receive and show the person parameters estimated by means of maximum likelihood (cf. Section 3.5), type

```
> theta <- person.parameter(rm_sum0)
> theta
```

```
Person Parameters:

 Raw Score      Estimate Std.Error
         1 -3.39916335 1.0848122
         2 -2.52177653 0.8317874
         3 -1.91597997 0.7365436
         4 -1.40925898 0.6923929
         5 -0.94517977 0.6731138
```

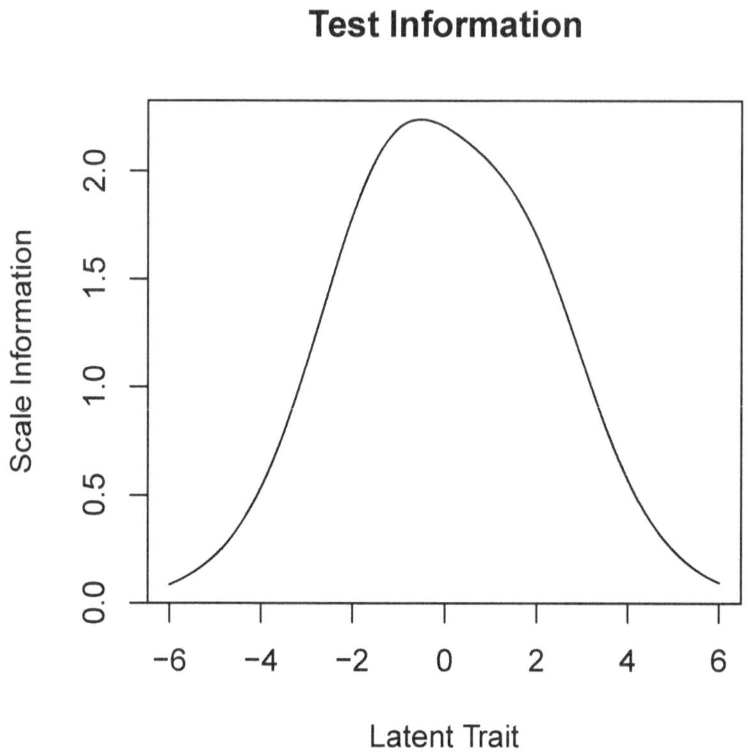

FIGURE 6.9: Test information curve.

```
 6 -0.49649948 0.6684716
 7 -0.04741163 0.6729622
 8  0.41178321 0.6830593
 9  0.88780184 0.6976852
10  1.38917318 0.7203110
11  1.93510329 0.7620310
12  2.57656204 0.8512971
13  3.48323587 1.0967887
14  4.45486143        NA
```

Note that this table does not show an estimate for every person. Recall from Chapter 2 that a person's ability estimate only depends on the number of items that they correctly answered. This means that we only need to compute an ability estimate for each possible sum score (termed `raw scores` in the table), and we can assign that estimate to each person earning that score. For example, we would estimate the ability of any person correctly answering ten items to be about 1.39.

The ability estimates are largely intuitive. First, we see that ability estimates increase with raw score. This makes sense, since a test taker is more likely to correctly answer an item when his or her ability exceeds the difficulty of that item. The more items correctly answered, the higher a test taker's ability is likely to be. Additionally, we see that the standard error increases with distance from zero, as was to be expected from the person item map or the item and test information curves, where we saw that most items are located around zero.

What may at first sight look surprising is the lack of a standard error for test takers correctly responding to all 14 items. The reason is that no maximum likelihood estimate exists for this perfect score, as discussed in Section 3.5. To deal with this, the `person.parameter()` function uses a method called spline interpolation to produce an ability estimate, but the procedure does not give error estimates. The same would be true for test takers solving 0 items correctly, but in this sample a score of zero did not occur.

We can get information about the ability estimates of individual test takers using the `summary()` function, i.e.,

```
> summary(theta)


Estimation of Ability Parameters

Collapsed log-likelihood: -76.27866
Number of iterations: 10
Number of parameters: 13

ML estimated ability parameters (without spline interpolated
 values):
             Estimate Std. Err.      2.5 %       97.5 %
theta 1   -0.49649948 0.6684716 -1.8066798  0.81368087
theta 2   -1.40925898 0.6923929 -2.7663241 -0.05219381
theta 3   -1.40925898 0.6923929 -2.7663241 -0.05219381
theta 4   -0.49649948 0.6684716 -1.8066798  0.81368087
theta 5    0.41178321 0.6830593 -0.9269884  1.75055486
theta 6   -0.94517977 0.6731138 -2.2644586  0.37409909
theta 7   -0.94517977 0.6731138 -2.2644586  0.37409909
theta 8   -0.04741163 0.6729622 -1.3663933  1.27157002
theta 9   -1.91597997 0.7365436 -3.3595790 -0.47238097
theta 10  -0.49649948 0.6684716 -1.8066798  0.81368087
```

The output – which we have again truncated here to save space – contains estimates and standard errors, together with the lower (2.5%) and upper (97.5%) boundaries of 95% confidence intervals, for all test takers answering up to 13 items correctly. Those test takers correctly answering all 14 items (or no item) are omitted from this output. We can also see that some test takers, for example the second and third, receive exactly the same ability and standard error. This is due to the fact that they have worked on the same set of items and scored the same sum score. Alternatively, we can use the command `coef(theta)` to obtain just the ability estimate for each test taker.

6.5 Test Evaluation in Small Data Sets

The **eRm** package also offers some nonparametric goodness-of-fit tests that allow to check the fit in small data sets. The theoretical foundation of these tests was explained in Section 4.3.7. Since they are based on bootstrapping, they cannot be applied to large data sets for computational reasons. We therefore demonstrate their application with a small artificial data set, which is part of the **eRm** package:

```
> data("raschdat1", package = "eRm")
```

This loads a response matrix named `raschdat1`, which consists of the simulated responses of 100 respondents to 30 items. We now illustrate the application of an overall goodness-of-fit test and the calculation of a goodness-of-fit test for item pairs. We first apply the overall goodness-of-fit test, which is based on the T_{11} statistic. The corresponding command is `NPtest`, and it is used as follows

```
> NPtest(as.matrix(raschdat1), method = "T11")


Nonparametric RM model test: T11 (global test - local
    dependence)
(sum of deviations between observed and expected inter-item
    correlations)
Number of sampled matrices: 500
one-sided p-value: 0.594
```

Here the response matrix `raschdat1`, which is internally stored as a data frame, is first transformed into a matrix for compatibility reasons. The second argument is the statistic `T11`, which should be applied. As was outlined in Section 4.3.7, these nonparametric tests are based on generating several artificial data sets, and by default 500 are used in **eRm**. This is also visible in the output. The output further contains a p-value for the model test. A p-value below 0.05 would indicate a significant violation of the Rasch model, but this is not the case here.

An alternative test statistic is Q_{3h}, which checks the model fit of item pairs. For Q_{3h}, small p-values indicate that two items show a *h*igher correlation than predicted by the Rasch model. Later we will also look at Q_{3l}, which indicates whether two items show a *l*ower correlation than predicted. To calculate Q_{3h} for every item pair, the following command can be applied:

```
> NPtest(as.matrix(raschdat1), method = "Q3h")
```

As was outlined above, this method is based on a random number generation. To make the output reproducible, we could set a seed, as was also recommended in Section 5.2. This can be done by the following command:

```
> NPtest(as.matrix(raschdat1), method = "Q3h", seed = 3)
```

Here (and in the `set.seed()` command), the `seed` can be any number from 1 to 2147483647, to make the outcome reproducible. In case that this argument is 0, a random number is chosen for the seed argument. The object created by `NPtest` consists of a list with the following elements:

- `n_eff`: The number of artificial data sets (or *eff*ective matrices) used for calculating the p-values, that is, the number of bootstrap samples.
- `prop`: A vector of all p-values calculated for the data set; for instance, in the case of Q_{3h} and Q_{3l}, it is a vector that contains the p-values for all item pairs.
- `Q3hmat`: A matrix of all p-values for the individual item pairs. In general, this is the most relevant part of the output.

Instead of showing the entire list, we create a more concise output in the following. First, we store the matrix of p-values under a new name

```
> Q3h_output <- NPtest(as.matrix(raschdat1),
+                        method = "Q3h", seed = 3)$Q3hmat
```

We can now easily access the individual rows and columns of this matrix of p-values. For instance, the p-values for all item pairs for the first nine items can be obtained by:

```
> Q3h_output[1:9,1:9]
      [,1]  [,2]  [,3]  [,4]  [,5]  [,6]  [,7]  [,8] [,9]
 [1,]    NA    NA    NA    NA    NA    NA    NA    NA   NA
 [2,] 0.438    NA    NA    NA    NA    NA    NA    NA   NA
 [3,] 0.130 0.662    NA    NA    NA    NA    NA    NA   NA
 [4,] 0.222 0.438 0.266    NA    NA    NA    NA    NA   NA
 [5,] 0.994 0.136 0.210 0.804    NA    NA    NA    NA   NA
```

```
 [6,] 0.142 0.518 0.516 0.846 0.944    NA    NA    NA   NA
 [7,] 0.994 0.426 0.628 0.514 0.396 0.948    NA    NA   NA
 [8,] 0.752 0.524 0.410 0.310 0.132 0.510 0.532    NA   NA
 [9,] 0.018 0.074 0.174 0.294 0.558 0.372 0.804 0.802   NA
```

Note that this matrix is symmetric but only the bottom left triangle is displayed.

Small p-values for Q_{3h} indicate that the corresponding two items show a higher correlation in their residuals than expected under the Rasch model. To find out for the entire matrix which item pairs show significant p-values, we can use the following command:

```
> which(Q3h_output < 0.05, arr.ind=TRUE)
      row col
 [1,]   9   1
 [2,]  26   1
 [3,]  18   2
 [4,]  12   4
 [5,]  26   4
 [6,]  24   6
 [7,]  13   8
 [8,]  16   9
 [9,]  12  10
[10,]  24  10
[11,]  12  11
[12,]  21  11
[13,]  22  14
[14,]  27  16
[15,]  28  18
[16,]  23  19
[17,]  26  19
[18,]  24  20
[19,]  22  21
[20,]  29  25
```

This summary shows that several item pairs show higher correlations in their residuals than expected, including item 9 with item 1, item 26 with item 1, etc.

The analogous command for computing the Q_{3l} statistic for each item pair is

```
> NPtest(as.matrix(raschdat1), method = "Q3l")
```

For Q_{3l}, small p-values indicate that the corresponding two items show a lower correlation than expected under the Rasch model. (We omit the results to save space, but the same R commands as above can be used for processing them.)

Since these statistics are calculated for all item pairs, using them on large sets of items is likely to lead to false-positive results. On the other hand, adjusting the *p*-values by means of, e.g., a Bonferroni correction to avoid false-positive results will be too strict if the number of items is high. We therefore recommend defining which item pairs should be checked before the analysis. For instance, an inspection of the item content could reveal that two items have very similar content, which could lead to the plausible hypothesis that test takers would give overly similar responses to these two items, or that responding to the first item could help with responding to the second item via a learning effect. Therefore, it would make sense to inspect the Q_{3h} statistic and its *p*-value only for certain item pairs, and not for all items. A similar recommendation was given in Debelak and Koller (2020).

6.6 Exercises

1. Bring the data set `SPISA` from the **`psychotree`** package into the workspace. Get the responses for items 37 through 45. These are the responses to the natural science items. Store them under the name `natSci_responses`. (This has already been done in the exercises for Chapter 5.) Store the gender information from `SPISA` into a vector named `gender`.
2. (Install and) load the **`eRm`** package. Fit a Rasch model to the `natSci_responses`. Summarize the output and plot the expected ICCs for all items. Which items are the easiest and the most difficult?
3. Create a person item map to see whether persons and items are well aligned.
4. Perform a likelihood ratio test to determine whether the item parameter estimates differ between test takers of different genders.
5. Use the graphical test in **`eRm`** to assess whether any individual items exhibit DIF with respect to gender. Which items appear to be easier for which gender?
6. Use the Wald test in **`eRm`** to assess again whether any individual items exhibit DIF with respect to gender.
7. Use the `stepwiseIT()` function in **`eRm`** to eliminate one item at a time.

   (Install and) load the package **`psychotools`** and use the `anchortest()` function to conduct a Wald test for gender with the iterative `forward` mean test statistic threshold (`MTT`) anchor item selection approach.

Do the results from the graphical and Wald test replicate when you use stepwise purification or anchoring?

8. Calculate infit and outfit statistics for all items in this data set. Which items show t values above 2 or below -2? Plot the empirical ICCs for these items.
9. Estimate and print the person parameters. Why do some persons receive the same estimate, while for other persons no estimate is printed by the `summary()` function?

7

R Package `mirt`

CONTENTS

The `mirt` package (Chalmers, 2021) contains functionality for fitting a variety of IRT models – including the Rasch, 2PL, and 3PL models as well as several models for polytomous responses (cf. Chapter 11) – using marginal maximum likelihood (cf. Section 3.3).

You will notice that in this chapter we will look at the FIMS data more from a modeling point of view. In the previous chapter, we have seen that some FIMS items show empirical ICCs that are not well in line with the Rasch model. Therefore, in this chapter we will estimate a Rasch, a 2PL, and a 3PL model, and select the model that best fits the data. Later we will look at the different options `mirt` offers for estimating the person parameters and assessing the fit of individual items.

Before beginning, make sure that you have installed the `mirt` package and loaded it into the workspace. If you have not yet installed `mirt`, type

```
> install.packages("mirt")
```

to install it. You can then load into the workspace by typing

```
> library("mirt")
```

We return to the data set consisting of the first 400 FIMS test takers that we already worked with in the previous chapters. Please make sure to repeat the steps for loading and preparing the data set in case you have started a new R session in the meantime.

```
> data(data.fims.Aus.Jpn.scored, package = "TAM")
> people <- 1 : 400
> responses <- data.fims.Aus.Jpn.scored[people, 2:15]
> colnames(responses) <- gsub("M1PTI", "I", colnames(responses))
```

7.1 Model Selection

We can estimate the item parameters of a Rasch model from this data set using the `mirt()` function by entering

```
> mirt_rm <- mirt(responses, 1, "Rasch")
```

The first argument to `mirt()` specifies the data. The second argument specifies the number of dimensions for our – potentially multidimensional – IRT model. Since we want to fit a unidimensional Rasch model, we set it to 1. The third argument specifies the IRT model we want to use. This argument is a single string, like `"Rasch"` or `"2PL"`. For a complete listing of the arguments to `mirt()`, see its help file.

We can also estimate the 2PL and 3PL models using `mirt()` by replacing `"Rasch"` with either `"2PL"` or `"3PL"` in the previous line of code.

```
> mirt_2pl <- mirt(responses, 1, "2PL")
> mirt_3pl <- mirt(responses, 1, "3PL")
```

We note at this point that the underlying estimation procedure of `mirt` is iterative. In a non-technical language, this means that the estimates are improved in a stepwise process until the estimation converges or until a maximum number of steps is reached. If this maximum number of steps is reached before the estimates converge, the item parameter estimates are considered unstable and `mirt` does not provide them. Instead, an error message such as `EM cycles terminated after 500 iterations` is shown. In this case, it is advisable to repeat the analysis after increasing the maximum number of steps by using the following additional argument in the `mirt()` function: `technical = list(NCYCLES = 2000)`. Here, the new maximum number of steps would be 2000, but of course you could also try it with another larger number.

We can use the different models to assess different assumptions of the Rasch model. For example, we can check whether every item has the same discrimination by comparing the Rasch model with the 2PL. In `mirt` model comparisons between fitted IRT models are performed using the `anova()`

function, which computes several information criteria (cf. Section 4.5.3) as well as a likelihood ratio test (cf. Section 4.5.2).[1]

We perform a model comparison by typing

```
> anova(mirt_rm, mirt_2pl)

Model 1: mirt(data = responses, model = 1, itemtype = "Rasch")
Model 2: mirt(data = responses, model = 1, itemtype = "2PL")
       AIC    SABIC       HQ      BIC   logLik     X2  df   p
1 5662.973 5675.249 5686.683 5722.845 -2816.487    NaN NaN NaN
2 5575.203 5598.118 5619.462 5686.964 -2759.601 113.77  13   0
```

The last four columns are relevant to the likelihood ratio test. The first of these, `logLik`, displays the maximum value of the log-likelihood for both models. The log-likelihood for the Rasch model, labelled "Model 1", is given in the first row, and the log-likelihood for the 2PL model, labelled "Model 2", is given in the second row. The next two columns, `X2` and `df`, display the χ^2 test statistic of the likelihood ratio test and its degrees of freedom. The final column, `p`, displays the p-value. Since there is only one value for the χ^2 statistic, degrees of freedom, and p-value for the comparison of the two models, the values of these columns are set to `NaN` in the first row. (`NaN` is shorthand for "not a number".)

The table lists the (rounded) p-value as 0. This indicates that the 2PL provides a significantly better fit than the Rasch model, suggesting that the items do not all have the same slope.

We can also compare the Rasch model and 2PL model using information criteria. The result of `anova()` includes three widely used information criteria: AIC, SABIC, and BIC. These are the Akaike Information Criterion (AIC), the Bayesian Information Criterion (BIC), and its sample size adjusted counterpart (SABIC). When using information criteria, we select the model with the lowest value. For the FIMS data, all information criteria agree and show a smaller value for the 2PL model in the second row. This indicates again that the 2PL provides a better account of the data than the Rasch model, even after penalizing for the additional item parameters.

We can similarly compare the 2PL and 3PL models using the `anova()` function.

```
> anova(mirt_2pl, mirt_3pl)

Model 1: mirt(data = responses, model = 1, itemtype = "2PL")
Model 2: mirt(data = responses, model = 1, itemtype = "3PL")
```

[1]As an alternative, the `mirt` package also provides a parametric bootstrap likelihood ratio test in the `boot.LR` function. Like the standard likelihood ratio test, this test is only suited for comparing nested models.

```
       AIC    SABIC       HQ      BIC    logLik     X2  df   p
1 5575.203 5598.118 5619.462 5686.964 -2759.601    NaN NaN NaN
2 5564.419 5598.792 5630.808 5732.061 -2740.210 38.784  14   0
```

We have to be careful at this point. The 2PL model results from the 3PL model by setting a specific type of parameter, namely the guessing parameter to a value of 0. Since the guessing parameter can take on values between 0 and 1, this means that the restricted model in our analysis, which is the 2PL model, results from the more general model by setting one parameter to a value on the edge of its parameter space. In this scenario, the classical likelihood ratio test can lead to inaccurate results, as was discussed in Section 4.5.2.[2] Therefore, we use only the information indices to compare the 2PL and the 3PL model.

The AIC suggests that the 3PL provides an even better account of the test responses than the 2PL. This may be due to guessing in one or more of the test items. The BIC-based information criteria, on the other hand, show a slight preference for the sparser 2PL model. As discussed in Section 4.5.3, this is due to the fact that in the BIC the number of model parameters is more strongly penalized.

7.2 Item Parameter Estimates

7.2.1 Illustration via Expected ICCs

We can better understand why the 3PL is preferred by some of the information criteria by viewing the expected ICCs for each item. Their shape is based on the estimated item parameters for difficulty, slope, and guessing. The expected ICCs can be drawn using the `plot()` function with `type = "trace"` ("trace lines" are an alternative name for ICCs). The resulting plot is shown in Figure 7.1.

```
> plot(mirt_3pl, type = "trace")
```

We can see that item discrimination and guessing vary considerably across items. Let us consider a few of those items that already caught our attention in the previous chapter. For example, in Figure 6.2 we found that item 19 had a steeper slope than was expected under the Rasch model. Accordingly, item 19 is among the items with a steeper slope in Figure 7.1, now that we have allowed the slopes to vary by using the 3PL model.

[2]This was not the case for the comparison of the Rasch model and the 2PL model. In this comparison, the restriction of the slope parameters to 1 is not at the boundary of the parameter space, because slope parameters can in principle take arbitrarily small or large values.

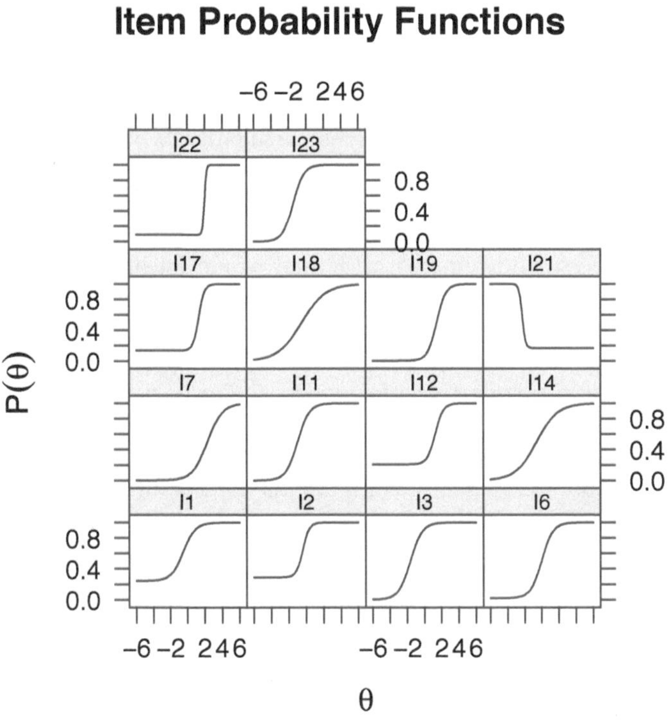

FIGURE 7.1: ICCs estimated by the 3PL model.

For item 12, the empirical ICC in Figure 6.2 indicated that guessing could be an issue. Again, we find in Figure 7.1 that – now that we have allowed this by using the 3PL model – item 12 does indeed show a lower asymptote above zero.

Interestingly, item 21, whose empirical ICC in Figure 6.2 was very flat, even shows an inverted pattern with a negative slope in Figure 7.1, now that the slope can be freely estimated for each item. This also demonstrates that the `mirt()` function does not constrain the slopes to be positive by default. The reason for this is that negative slopes are commonplace in multidimensional IRT models. A negative slope indicates that, all else being equal, a person with a higher value on the latent trait is less likely to give a positive response than a person with a lower value.

What a negative slope estimate means for test construction depends on context. For achievement tests, a negative slope indicates that a test item is not behaving like it should. This could be the result of, e.g., differential item functioning. Item 21 showed a significant violation of the Rasch model in Andersen's likelihood ratio test with the mean sum score as the split criterion in Section 6.2.3, and its empirical ICC in Figure 6.2 included a particularly low value for high abilities. When a negative slope occurs, it should also be double-checked whether perhaps the item was graded incorrectly. However, as we will show below, in this example we should be particularly careful interpreting the negative slope, because the uncertainty of the estimate is very high.

For other psychological tests, such as personality or attitude scales, negative slopes may be intended for some of the items. In this setting, test designers routinely use negatively worded items. A negatively worded version of "I like long walks on the beach", for example, might be "I do not like long walks on the beach". A person agreeing with the first would ideally disagree with the second[3], so we could legitimately get a negative slope when the version we use is in the opposite direction to the rest of the items.

Another item that showed problems in the previous chapter is item 2. This item was also among those for which Andersen's likelihood ratio test with the mean sum score as the split criterion showed a significant violation of the Rasch model. This item also has a large guessing parameter (lower asymptote notably above zero) in Figure 7.1 for the 3PL. Guessing, which affects the probability of solving an item particularly for lower values of the person parameter, can be one reason why the item difficulty parameter in a Rasch model differs between test takers with below vs. above average abilities.

For yet other items the pattern is less clear. For example, item 22 also showed a significant violation of the Rasch model in the graphical model test with the mean sum score as the split criterion in Section 6.2.3, which could indicate a slope that is larger or smaller than those of the other items. In addition, it showed a value of over 2 for the outfit t statistic (and over 1.3 for the outfit MSQ) in Section 6.2.8. This by itself would suggest underfit. The infit t statistic, however, which is less affected by the residuals of persons located further away from the item difficulty, shows a small but negative value for this item, which would point in the opposite direction. In Figure 7.1 for the 3PL fit, item 22 shows a very steep slope, but also a certain degree of guessing.

We do not hypothesize about possible causes for these findings, but the interested reader can look up the items by following the link provided in Chapter 5. However, before rushing to conclusions, note that we will discuss the uncertainty of the point estimates, in particular for the slopes of items 21 and 22, below.

7.2.2 Displaying the Estimates

We can display the item parameter estimates for the 3PL using the `coef()` function. Recall that we also used the `coef()` function to obtain the fitted values after estimating the Rasch model parameters using the `RM()` function in the `eRm` package. In R, the `coef()` function is a generic function. This means

[3]Note, however, that negatively worded items can work differently beyond being coded in the opposite direction. It is recommended to include a method factor to account for this (cf., e.g., Weijters, Baumgartner, & Schillewaert, 2013).

that the output of the function depends on how its argument was generated. For example, an object containing the result of a call to `mirt()` will be treated differently than an object containing the result of a call to `RM()`. This allows package developers to provide a consistent interface across R packages.

In this case, we use the `coef()` function as follows.

```
> coef(mirt_3pl, IRTpars = TRUE, simplify = TRUE)
$items
          a      b     g u
I1    1.410 -0.493 0.244 1
I2    2.667 -0.404 0.291 1
I3    1.445 -1.633 0.000 1
I6    1.514  0.009 0.019 1
I7    1.163  2.297 0.000 1
I11   1.487 -0.984 0.000 1
I12   2.131  1.275 0.207 1
I14   0.805 -0.586 0.000 1
I17   3.076  1.320 0.138 1
I18   0.724 -0.510 0.000 1
I19   1.844  1.449 0.000 1
I21  -5.264 -2.271 0.167 1
I22   9.179  1.990 0.090 1
I23   1.625 -1.452 0.000 1

$means
F1
 0

$cov
   F1
F1  1
```

The result is a list of three elements: `items`, `means`, and `cov`. The latter two elements are the assumed mean and variance of the person ability distribution for the population of test takers. These are required, since we are using marginal maximum likelihood estimation. The first element, `items`, gives the item parameters. The slope or discrimination parameter is listed in the `a` column, the difficulty parameter in the `b` column, and the guessing parameter in the `g` column. The values of these estimates reflect the shapes we have seen for the expected ICCs in Figure 7.1. We see that many of the slope parameters in column `a` vary considerably from 1, and that many of the guessing parameters in column `g` vary considerably from 0.

Note that our last call of `coef()` has two arguments that were not used when calling `coef()` on the result of `RM()` in the previous chapter. The first of these arguments, `IRTpars`, is used to indicate whether we would like the results displayed using the "standard" parameterization of the 3PL model presented in Section 4.5.1.2. Under the hood, the `mirt()` function uses a "slope-intercept" form of the 3PL,

$$\Pr(U_{pi} = 1 \mid \theta_p, a_i, d_i, \gamma_i) = \gamma_i + (1 - \gamma_i) \cdot \frac{\exp\{a_i \cdot \theta_p + d_i\}}{1 + \exp\{a_i \cdot \theta_p + d_i\}}.$$

This form simplifies computations for multidimensional IRT models, but it is difficult to interpret for unidimensional IRT models. Omitting the argument `IRTpars = TRUE` or setting it to `IRTpars = FALSE` would lead to an adapted output in the slope-intercept form, with a column `d` instead of a column `b`. Setting `IRTpars = TRUE` tells the `coef()` function to convert its estimates from the slope-intercept form to the standard form from Section 4.5.1.2

$$\Pr(U_{pi} = 1 \mid \theta_p, \alpha_i, \beta_i, \gamma_i) = \gamma_i + (1 - \gamma_i) \cdot \frac{\exp\{\alpha_i \cdot (\theta_p - \beta_i)\}}{1 + \exp\{\alpha_i \cdot (\theta_p - \beta_i)\}}.$$

Here, α corresponds to `a`, β corresponds to `b` and γ corresponds to `g` in the output presented above. The second argument to the `coef()` function, `simplify`, indicates that we would like the output to be simplified. The `mirt()` function allows each item to have a different ICC, and these ICCs do not always have the same parameters. For example, we might want to model a test with dichotomous and polytomous items using the Rasch model for the dichotomous items and the Partial Credit model for the polytomous items. The dichotomous model has one difficulty parameter per item, while the polytomous model has multiple threshold parameters per item (for details see Section 11.1). To deal with this, the `coef()` function displays the items as a list by default. Recall from Chapter 5 that a list is an R data type for storing multiple pieces of related information with different types or sizes. When we use the same IRT model for every item, the output can be simplified into a `data.frame` and displayed like a table as shown above. This puts the results into a form that is easier to read.

However, in this output we see only the point estimates for the item parameters, and no information about their uncertainty. In order to compute standard errors and be able to display confidence intervals for each estimate, we need to refit the model with the option `SE = TRUE` and display the detailed instead of the simplified output

```
> mirt_3pl <- mirt(responses, 1, "3PL", SE = TRUE)
> coef(mirt_3pl, IRTpars = TRUE)


$I17

                a        b          g  u
par     3.0758348 1.319978 0.13808182  1
CI_2.5  0.8087457 1.011940 0.08277523 NA
```

```
CI_97.5 5.3429239 1.628017 0.19338840 NA

$I18
                a          b             g  u
par     0.7239776 -0.5103173  0.00004102312  1
CI_2.5  0.4369053 -0.8604131 -0.00790563438 NA
CI_97.5 1.0110499 -0.1602216  0.00798768062 NA

$I19
               a        b              g  u
par     1.843701 1.448607  0.000003537302  1
CI_2.5  1.134135 1.108365 -0.000743608230 NA
CI_97.5 2.553267 1.788848  0.000750682834 NA

$I21
                 a         b         g  u
par      -5.263578 -2.270616 0.1671759  1
CI_2.5  -25.534131 -3.032526 0.1260265 NA
CI_97.5  15.006975 -1.508705 0.2083253 NA

$I22
                 a        b          g  u
par       9.179159 1.989994 0.08967797  1
CI_2.5  -17.333347 1.579914 0.05909454 NA
CI_97.5  35.691665 2.400075 0.12026140 NA

$I23
               a         b             g  u
par     1.625240 -1.452268  0.00004396566  1
CI_2.5  1.047281 -1.804960 -0.00855664894 NA
CI_97.5 2.203199 -1.099576  0.00864458025 NA
```

Here we display only the output for a selection of items to save space. In this output, again we find entries for slope (`a`), difficulty (`b`), guessing (`g`), as well as for the upper asymptote of the 4PL (`u`), which is set to 1 for the 3PL. The output is organized by item, and provides the point estimate in the first row (`par`) together with the lower (`CI_2.5`) and upper (`CI_97.5`) boundary of a 95% confidence interval.

For most items, we find that the confidence interval boundaries for the slope `a` are both positive and rather close to each other, indicating that their positive slopes have been reliably estimated. However, when we inspect in more detail the slope estimates for items 21 and 22, which have caught our attention above, we see that their confidence interval boundaries span from approx. -25.5 to 15.0 for item 21 and from approx. -17.3 to 35.7 for item 22. These intervals are very wide, indicating a high degree of uncertainty, and also span both negative and positive values, indicating that the direction of the slope cannot be clearly determined for these items.

For some of the other item parameters we also find slightly wider confidence intervals, such as for the slope of item 17, which spans from approx. 0.8 to 5.3, but in this example it is most pronounced for the two slopes we have discussed. We generally recommend inspecting the standard errors or confidence intervals before interpreting any item parameters, because – as we have seen above – it is tempting to try and find an explanation for the shape of an item's ICC while it may be based on a highly uncertain point estimate.

7.3 Evaluating Goodness-of-Fit

The previous comparisons show that the 3PL provides a better account of the data than the Rasch or 2PL models. However, this does not tell us whether the account provided by the 3PL model is reasonable, simply that it is the most reasonable of the models that we compared. We can further test the fit of this model using the M_2 statistic (cf. Section 4.3.2), which is provided by the function `M2()` in `mirt`. To compute the M_2 statistic for the 3PL fit, enter

```
> M2(mirt_3pl)
            M2 df         p      RMSEA RMSEA_5   RMSEA_95
stats 76.06463 63 0.1249838 0.02279774       0 0.03943958
           SRMSR       TLI       CFI
stats 0.04533455 0.975068 0.9827394
```

The first three columns of the output give the test statistic, degrees of freedom and p-value for the M_2 statistic. In this case, the p-value tests the null hypothesis that the observed row and column sums are consistent with the 3PL model. In this case, the p-value is above 5%, meaning that the test provides no evidence that the 3PL cannot account for the observed data.

In addition to the M_2 statistic, the `M2()` function prints out the root mean square error of approximation (RMSEA, along with a 90% confidence interval) and the standardized root mean square residual (SRMR, both presented in Section 4.3.2), as well as two other fit indices from multivariate statistics based on the χ^2 test statistic. The `M2()` function computes each of these statistics, replacing χ^2 with its limited-information counterpart, M_2, as was explained in Section 4.3.1 and Section 4.3.2.

The most commonly reported of these statistics is RMSEA. A model is considered to give a good account of test data when the upper bound of the confidence interval falls below 0.05. In this case, the upper bound is ca. 0.039, suggesting that the model provides a good account of the data.

To obtain the itemwise $S\text{-}X_i^2$ statistic (cf. Section 4.3.4), one can use the following command

```
> itemfit(mirt_3pl)
   item    S_X2 df.S_X2 RMSEA.S_X2 p.S_X2
1    I1  4.732       6      0.000  0.579
2    I2  9.894       5      0.050  0.078
3    I3  5.328       6      0.000  0.503
4    I6  9.774       5      0.049  0.082
5    I7 10.211       5      0.051  0.069
6   I11 20.067       5      0.087  0.001
7   I12  8.406       7      0.022  0.298
8   I14 10.647       7      0.036  0.155
9   I17  8.728       6      0.034  0.189
10  I18  4.604       7      0.000  0.708
11  I19  7.320       4      0.046  0.120
12  I21  3.556       9      0.000  0.938
13  I22 15.248       7      0.054  0.033
14  I23  5.317       4      0.029  0.256
```

The resulting table consists of five columns. The first column contains the item labels, the second column presents the values of the $S\text{-}X_i^2$ statistics for all items, the third column their degrees of freedom and the fifth column their p-values under an assumed χ^2 distribution. The fourth column presents RMSEA statistics to evaluate the effect of a detected model violation. Large RMSEA values or small p-values indicate model violations.

We can see that most items show good fit for the 3PL model. Only items 11 and 22 show p-values below 0.05. Item 22 has already shown different peculiarities in this and the previous chapter. Item 11 has been diagnosed with overfit under the Rasch model in Section 6.2.8, and seems to still not be ideally described by the 3PL model. Again, we will not exclude any items here for comparability between the chapters, but in a real analysis, items that show misfit or other issues should be further investigated.

We can also obtain a number of other item fit statistics by means of the **fit_stats** argument to the **itemfit** function. For example,

```
> itemfit(mirt_3pl, fit_stats = "PV_Q1")
```

will compute the improved Q_1 statistic of Chalmers and Ng (2017, cf. Section 4.3.4). This command uses an analytical approach for the calculation of the p-values. The option **"PV_Q1*"**, instead of **"PV_Q1"**, additionally employs a bootstrap approach. This generally leads to more accurate p-values, but is also computationally a lot more intensive. It should be noted that both approaches use a plausible values imputation to account for the uncertainty in the estimation of the person parameters, which are used for the calculation of these statistics. As a consequence of this imputation, the outcome of this command usually varies slightly if it is carried out repeatedly. Like for any command involving random sampling, the results can be made reproducible by means of `set.seed()`:

```
> set.seed(1234)
> itemfit(mirt_3pl, fit_stats = "PV_Q1")
   item  PV_Q1 df.PV_Q1 RMSEA.PV_Q1 p.PV_Q1
1    I1  7.058    6.000       0.021   0.315
2    I2  6.412    4.000       0.039   0.170
3    I3  5.774    3.467       0.041   0.164
4    I6  8.957    7.000       0.026   0.256
5    I7  4.120    3.000       0.031   0.249
6   I11  6.986    5.300       0.028   0.251
7   I12  8.924    7.000       0.026   0.258
8   I14  9.732    7.000       0.031   0.204
9   I17 13.385    7.000       0.048   0.063
10  I18  8.254    7.000       0.021   0.311
11  I19  4.656    2.967       0.038   0.195
12  I21  9.636    7.000       0.031   0.210
13  I22 15.523    7.000       0.055   0.030
14  I23  5.172    3.067       0.041   0.166
```

We find that most items are well described by the 3PL model, only item 22 again shows significant misfit for Q_1.

Additional options for `fit_stats` are:

- `"Zh"` for the z_h statistic (Drasgow et al., 1985, cf. Section 4.3.6).
- `"X2"` for the Q_1 statistic of Yen (1981). By setting the additional arguments `group.fun = median` and `group.bins`, with the second argument being the desired number of respondent groups, one obtains the χ^2 statistic of Bock (1972) instead (cf. Section 4.3.4).
- `"G2"` for the G_i^2 statistic (cf. Section 4.3.4). Using the `group.bins` argument again, one can obtain the $Q_1 - G^2$ statistic of Orlando and Thissen (2000).

  Note again, as was already mentioned in Section 4.3.4, that especially the `X2` and `G2` statistics are not well suited for short tests.
- `"X2*"` and `"X2*_df"` for the fit statistics of Stone (2000, cf. Section 4.3.4). These statistics are computationally more intensive and typically require a longer computation time. Since these statistics use simulation experiments under the hood, we also recommend setting a seed here to make the outcome reproducible.
- and `"infit"` for the infit and outfit MSQ and t statistics (the latter being termed `z` here, cf. Section 4.3.3).

When computing different fit statistics, the reader should also be reminded that they are constructed quite differently, as we have outlined in Section 4.3. Accordingly, they are sensitive to different patterns of misfit. This means that different fit statistics will not necessarily flag the same items.

Either of these statistics can be used to investigate any misfit that remains when using the 3PL model, as we have shown above. However, of course we can also use the `itemfit` function in `mirt` to compute additional fit statistics for Rasch or 2PL models. For this we only need a fitted Rasch or 2PL model object, that has been created by `mirt`, such as `mirt_rm` or `mirt_2pl`. We can then supply this model object instead of `mirt_3pl` to the itemfit function.

Note that when we do assess the fit for the Rasch model by means of, e.g., infit and outfit statistics (cf. Section 4.3.3), which are available for the Rasch model in both `eRm` and `mirt`, it is again possible that the different packages flag somewhat different sets of items for misfit. The reason for this is that the two packages use different methods for estimating the item parameters (conditional vs. marginal maximum likelihood), but also for estimating the person parameters. By using maximum likelihood estimators for the person parameters in both packages (instead of EAP estimators, which are the default in `mirt`, as we show in the next section), we get results that are similar to those from `eRm`.

```
> itemfit(mirt_rm, fit_stats = "infit", method = "ML")
   item outfit z.outfit infit z.infit
1    I1  0.814   -1.683 0.904  -1.665
2    I2  0.685   -2.793 0.809  -3.271
3    I3  0.724   -1.510 0.870  -1.493
4    I6  0.834   -2.272 0.860  -3.258
5    I7  0.682   -1.358 0.831  -1.504
6   I11  0.832   -1.381 0.816  -3.143
7   I12  1.148    1.484 0.971  -0.554
8   I14  0.990   -0.086 1.023   0.494
9   I17  1.315    2.276 0.935  -1.040
10  I18  1.083    1.011 1.019   0.426
11  I19  0.569   -2.585 0.748  -3.014
12  I21  2.279    5.852 1.244   2.960
13  I22  1.823    2.911 0.978  -0.165
14  I23  0.690   -1.831 0.853  -1.804
```

If we employ the same rule of thumb and flag items with an infit or outfit `z` value (which in Section 4.3.3 and Section 6.2.8 was called t) below -2 or above 2, we find that for items 2, 6, 11, and 19, at least one value is below -2, which indicates overfit. For items 17, 21, and 22, at least one value is above 2, which indicates underfit. These results are in line with those from eRm in Section 6.2.8. We can also compute the Q_3 statistic from Section 4.3.5 for item pairs based on the Rasch model:

```
> Q3_mat <- residuals(mirt_rm, type="Q3")
> which(abs(Q3_mat) > 0.2, arr.ind=TRUE)
    row col
I1    1   1
I2    2   2
I3    3   3
I6    4   4
I7    5   5
I11   6   6
I12   7   7
I14   8   8
I17   9   9
I18  10  10
I19  11  11
I21  12  12
I22  13  13
I23  14  14
```

We find that no item pair exceeds the rule of thumb threshold[4] of 0.2 mentioned (among other options, cf. Section 4.3.5) by Christensen et al. (2017) – all listed pairs are for correlations of an item with itself, which of course are equal to one by definition.

The mirt package also offers a function for detecting DIF items based on multiple group models. The DIF() function can be used with preassigned anchor items and also with a sequential procedure. For details see the examples for DIF() in the mirt help pages.

7.4 Ability Estimation

We can estimate test taker ability using the fscores() function. For unidimensional models, the most important arguments to fscores() are object and method. The object argument takes the result of the mirt() function.

The method argument indicates which method to use for estimating the person parameters. By default, method="EAP", indicating that the person

[4]Note that – while in eRm p-values were available for a Q_3-based statistic and related statistics through a nonparametric test framework for small data sets – here we inspect the values of the statistics themselves. This is why here we compare the values with the rule of thumb threshold of 0.2 for the descriptive statistics, rather than with the typical 0.05 for significance tests, which was used in a similar command in Section 6.5.

parameter should be estimated using the expected a posteriori (EAP) estimator. We can compute the EAP estimates for the 3PL model and print out its first six entries by entering

```
> theta_eap <- fscores(mirt_3pl)
> head(theta_eap)
             F1
[1,] -0.2781089
[2,] -0.6800843
[3,] -0.8095114
[4,] -0.1274342
[5,]  0.4866075
[6,] -1.0577696
```

Again, by default `mirt` displays only the point estimates, but it is possible to add standard errors by means of the option `full.scores.SE = TRUE` to the `fscores()` function. The standard errors should be inspected before interpreting or reporting person parameter estimates.

The `fscores()` function also provides maximum likelihood (ML), maximum a posteriori (MAP) and weighted likelihood (WLE) estimators. All these approaches for estimating the person parameters after the item parameters have already been estimated by means of marginal maximum likelihood have been discussed in Section 3.5.

We now compare the four types of person parameter estimates provided by `mirt`. Please note that we do this only for didactic reasons. In IRT analyses of empirical data, researchers typically only report one type of estimate of the parameters.

The ML, MAP, and WLE estimators can be computed by entering

```
> theta_ml <- fscores(mirt_3pl, method = "ML", max_theta = 30)
> theta_map <- fscores(mirt_3pl, method = "MAP")
> theta_wle <- fscores(mirt_3pl, method = "WLE")
```

For the maximum likelihood estimation, by setting the `max_theta` argument to 30 we allow person parameter estimates to lie between −30 and 30. By default `mirt` assumes that all person parameters lie between −20 and 20, but for this data set even more extreme maximum likelihood estimates result for some persons, so that the default setting would lead to a warning.

We can then compare the results for the different estimators by means of the following lines of code.

```
> ests <- cbind(theta_eap, theta_ml, theta_map, theta_wle)
> colnames(ests) <- c("EAP", "ML", "MAP", "WLE")
> pairs(ests, xlim = c(-3, 3), ylim = c(-3, 3))
```

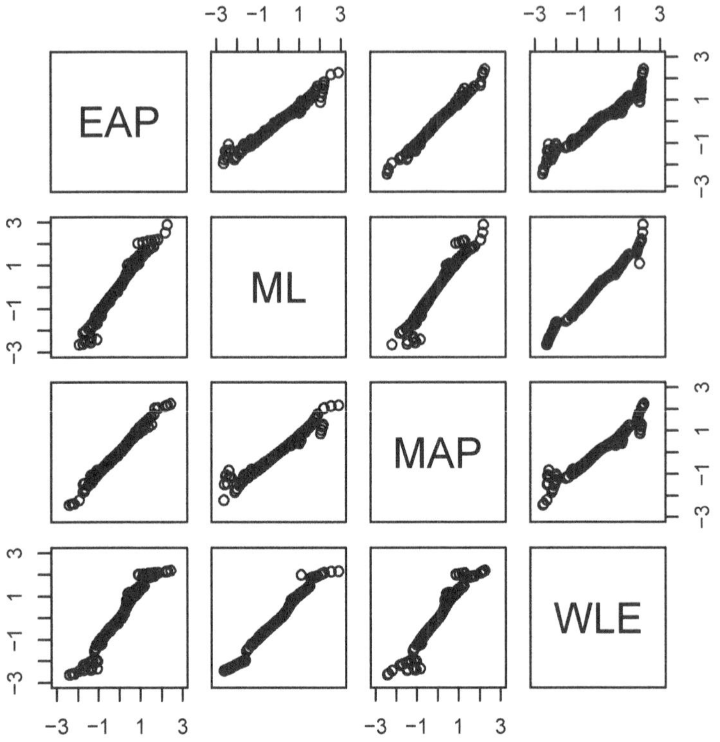

FIGURE 7.2: Scatterplots for comparing different estimators.

The results are shown in the scatterplots in Figure 7.2. Overall, the estimates are similar (i.e., close to a straight diagonal line) in the center of each plot. However, differences are notable for estimates further away from the center of the latent continuum, as was to be expected from Section 3.5.[5]

The `mirt` package also offers the z_h person fit statistic, which we have discussed in Section 4.3.6, via the `personfit()` function.

[5]For the maximum likelihood estimates, two types of values are not shown in Figure 7.2: Persons with a perfect score or a score of zero receive a value of plus or minus infinity, as was explained in Section 3.5. Additionally, for persons with an almost perfect or zero score the log-likelihood may not have a unique maximum, resulting in extreme values of plus or minus 20 or beyond. We allowed this by setting the `max_theta` argument to 30, but such extreme maximum likelihood estimates may indicate there is not enough evidence from the item responses to estimate the person parameters by means of maximum likelihood, while the other estimation approaches additionally rely on Bayesian prior information.

7.5 Exercises

1. Bring the data set `SPISA` from the `psychotree` package into the workspace. Get the responses for items 37 through 45. These are the responses to the natural science items. Store them under the name `natSci_responses`. (This has already been done in the exercises for the previous chapters.) Assign column names to the item response matrix by means of the following command (`mirt` will not accept items without names):

```
> colnames(natSci_responses) <- paste("I", 1:9)
```

2. (Install and) load the `mirt` package. Fit the `natSci_responses` using a Rasch model, a 2PL and a 3PL model, and store the results.
3. Determine which model provides the best account of the observed responses. Use this model in the following exercises.
4. Plot the expected ICCs. Are the results in line with the in/outfit statistics and empirical ICCs from the exercises for the previous chapter?
5. Display the item parameter estimates together with their confidence intervals. Which items have the lowest and highest discrimination parameters?
6. Compute M_2, $S\text{-}X_i^2$, and the fit statistic of Stone (2000) to assess the item fit for the 2PL model.
7. Estimate the person parameters by means of the maximum a posteriori approach and display the estimates for the first few persons, together with their standard errors.

8

R Package TAM

CONTENTS

In this chapter, we will demonstrate how the TAM package is used. We will demonstrate the central functions for estimating item and person parameters as well as for assessing goodness of fit. To avoid repetition and keep it simple, for this and the following chapter we will return to the Rasch model for illustrating the general usage of the software. We would like to emphasize that we restrict ourselves to the Rasch model here for didactic reasons, but that the TAM package offers a multitude of functionalities beyond those presented here, including a latent class approach proposed by Formann (1992). To find out more about the options provided by TAM, please consult its help pages.

If you have been following along, you will already have the TAM package installed, because it provides the FIMS data set we have been using in the previous chapters. Otherwise, you can install it by entering

```
> install.packages("TAM")
```

Once the package has been installed, load it into the workspace by typing

```
> library("TAM")
```

Again, we will use the data set consisting of the first 400 FIMS test takers. Please make sure to repeat the steps for loading and preparing the data set in case you have started a new R session in the meantime.

```
> data(data.fims.Aus.Jpn.scored, package = "TAM")
> people <- 1 : 400
> responses <- data.fims.Aus.Jpn.scored[people, 2:15]
> colnames(responses) <- gsub("M1PTI", "I", colnames(responses))
```

8.1 Item Parameter Estimation

We start our demonstration of the TAM package with fitting the Rasch model.

```
> tam_rm <- tam.mml(responses)
```

Like mirt, the TAM package uses marginal maximum likelihood estimation, which was outlined in Section 3.3. TAM also offers functions for fitting other IRT models via marginal maximum likelihood, for example, tam.mml.2pl() for fitting the 2PL model.

If you use this function, note that TAM also uses a parameterization similar to the "slope-intercept" form we have seen for mirt in Section 7.2.2, which differs from the standard IRT parameterization presented in Section 4.5.1.1:

$$\Pr(U_{pi} = 1 \mid \theta_p, \xi_i, B_i) = \frac{\exp\{B_i \cdot \theta_p - \xi_i\}}{1 + \exp\{B_i \cdot \theta_p - \xi_i\}}. \tag{8.1}$$

Here, B_i is a discrimination or slope parameter, whereas ξ_i can be seen as an item difficulty parameter that also incorporates the item discrimination. The Rasch model is obtained by setting B_i to 1. By comparing this function with the item response function given by Equation (4.1) in Section 4.5.1.1, we see that B_i corresponds to α_i, whereas ξ_i corresponds to $\alpha_i \cdot \beta_i$. Large positive values of ξ_i thus typically indicate difficult items with a large discrimination parameter.

We can inspect the item parameters via

```
> tam_rm$item
    item   N      M    xsi.item  AXsi_.Cat1 B.Cat1.Dim1
I1    I1 400 0.7175 -1.10292502 -1.10292502           1
I2    I2 400 0.7425 -1.25047869 -1.25047869           1
I3    I3 400 0.8525 -2.04219530 -2.04219530           1
I6    I6 400 0.5100 -0.04803217 -0.04803217           1
I7    I7 400 0.1000  2.53122857  2.53122857           1
I11  I11 400 0.7425 -1.25047869 -1.25047869           1
I12  I12 400 0.3350  0.81318838  0.81318838           1
I14  I14 400 0.6025 -0.49523230 -0.49523230           1
I17  I17 400 0.2450  1.32631928  1.32631928           1
I18  I18 400 0.5825 -0.39686510 -0.39686510           1
I19  I19 400 0.1450  2.06394307  2.06394307           1
I21  I21 400 0.1825  1.75454300  1.75454300           1
I22  I22 400 0.1100  2.41452138  2.41452138           1
I23  I23 400 0.8400 -1.93453524 -1.93453524           1
```

The columns of this output can be interpreted as follows:

- `item` indicates the name of the item.
- `N` indicates the number of test takers that responded to each item. In this case, all 400 test takers responded to each item.
- `M` is the mean of the responses to each item. An item with a high mean was correctly answered by a high percentage of test takers.
- `xsi.item` for the Rasch model is the item difficulty parameter. Items with high values tend to be more difficult. For example, item 7 is only answered correctly by 10% of the test takers (for dichotomous items, the mean response corresponds to the percentage of correct answers). Accordingly, it has a high difficulty. By contrast, item 3 is answered correctly by over 85% of the test takers and receives a low difficulty. Like in Chapter 6, we find that item 3 has the lowest and item 7 the highest difficulty.
- `AXsi_.Cat1` repeats the item difficulty for the Rasch model, but would allow the inclusion of a design matrix A, which we did not use here (see `TAM` help pages for details). For polytomous models (cf. Chapter 11), the output will include item parameters for more than one category.
- `B.Cat1.Dim1` is the item discrimination or slope parameter B_i from Equation (8.1). For the Rasch model, the slope is 1 for each item.

We can also display only the difficulty and its standard error by

```
> tam_rm$xsi
            xsi     se.xsi
I1  -1.10292502 0.1199213
I2  -1.25047869 0.1231185
I3  -2.04219530 0.1490275
I6  -0.04803217 0.1091832
I7   2.53122857 0.1742477
I11 -1.25047869 0.1231185
I12  0.81318838 0.1149197
I14 -0.49523230 0.1112816
I17  1.32631928 0.1249697
I18 -0.39686510 0.1105217
I19  2.06394307 0.1501087
I21  1.75454300 0.1378399
I22  2.41452138 0.1675220
I23 -1.93453524 0.1445361
```

where the `xsi` column contains the item difficulty, followed by its standard error in column `se.xsi`.

Finally, we can get the IRT parameters in the classical parameterization using:

```
> tam_rm$item_irt
   item alpha        beta
1    I1     1 -1.10292502
2    I2     1 -1.25047869
3    I3     1 -2.04219530
4    I6     1 -0.04803217
5    I7     1  2.53122857
6   I11     1 -1.25047869
7   I12     1  0.81318838
8   I14     1 -0.49523230
9   I17     1  1.32631928
10  I18     1 -0.39686510
11  I19     1  2.06394307
12  I21     1  1.75454300
13  I22     1  2.41452138
14  I23     1 -1.93453524
```

Here, the column `alpha` contains the item discrimination parameters, which are 1 in the Rasch model, and the columns `beta` contains the item difficulty parameters.

8.2 Evaluating Goodness-of-Fit

The `TAM` package offers various options for assessing the fit of an estimated IRT model. For example, the following command computes several goodness-of-fit statistics, many of them for item pairs:

```
> rm_gof <- IRT.modelfit(tam_rm)
```

The resulting object is a list, which provides the various fit statistics. Here we will only list a few that we have treated in Chapter 4.

- `rm_gof$chi2.stat` provides the X^2 tests of Chen and Thissen (1997, cf. Section 4.3.5) for all item pairs, together with p-values with and without a Holm correction. The Holm correction increases the reported p-values to reduce the probability of false-positive results.
- `rm_gof$fitstat` gives three fit statistics, among them `SRMSR`. Values close to 0 indicate a good model fit (cf. Section 4.3.2).
- `rm_gof$residuals` contains the model residuals, which are the differences between the observed responses and their expected values (cf. Section 4.3.3).

- `rm_gof$Q3.matr` gives a matrix of the Q_3 statistics. Various cutoff values have been proposed for the absolute values of Q_3, for instance a rule of thumb threshold of 0.2 (cf. Section 4.3.5).

Here we present only some exemplary results from the extensive output of `IRT.modelfit`:

```
> rm_gof$fitstat
100*MADCOV        SRMR       SRMSR
1.30035167 0.07717775 0.09413315
```

This summary of global fit statistics contains the SRMSR (cf. Section 4.3.2). Its value of approx. 0.0941 for the Rasch model we have fit here exceeds the cutoff of 0.05 suggested by Maydeu-Olivares (2013).

Remember that we have shown in Chapter 7 that a 3PL model is better suited for describing the FIMS data, so it is not surprising that the Rasch model does not show good overall fit.

We can also inspect the Q_3 statistics for item pairs (cf. Section 4.3.5).

```
> Q3_mat <- rm_gof$Q3.matr
> which(abs(Q3_mat) > 0.2, arr.ind=TRUE)
     row col
```

We find that the output is empty, meaning that no item pair exceeds the rule of thumb threshold of 0.2. This corresponds to our results from Section 7.3.

It is also possible to compute in- and outfit statistics (cf. Section 4.3.3) in TAM

```
> tam.fit(tam_rm)
   parameter Outfit Outfit_t Outfit_p Outfit_pholm Infit Infit_t
1         I1  0.937   -1.139    0.255        1.000 0.962  -0.663
2         I2  0.837   -2.833    0.005        0.055 0.915  -1.417
3         I3  0.926   -0.796    0.426        1.000 0.982  -0.161
4         I6  0.920   -2.143    0.032        0.289 0.930  -1.860
5         I7  0.870   -1.074    0.283        1.000 0.963  -0.259
6        I11  0.923   -1.303    0.193        0.963 0.918  -1.367
7        I12  1.078    1.582    0.114        0.691 1.013   0.275
8        I14  1.048    1.103    0.270        1.000 1.038   0.888
9        I17  1.183    2.746    0.006        0.066 1.007   0.121
10       I18  1.070    1.662    0.097        0.691 1.042   0.999
11       I19  0.763   -2.680    0.007        0.074 0.901  -1.039
12       I21  1.714    7.330    0.000        0.000 1.192   2.268
13       I22  1.626    4.496    0.000        0.000 1.052   0.473
14       I23  0.854   -1.715    0.086        0.691 0.950  -0.540
```

```
   Infit_p Infit_pholm
1    0.507       1.000
2    0.157       1.000
3    0.872       1.000
4    0.063       0.817
5    0.796       1.000
6    0.172       1.000
7    0.783       1.000
8    0.374       1.000
9    0.904       1.000
10   0.318       1.000
11   0.299       1.000
12   0.023       0.327
13   0.636       1.000
14   0.589       1.000
```

for which we show an abbreviated output here to save space.

Again, we employ the rule of thumb from Section 4.3.3 and flag items with infit or outfit t values below -2 or above 2. We find that for items 2, 6, and 19, at least one value is below -2, but not for item 11, which was additionally flagged by `eRm` and `mirt`. For items 17, 21, and 22, at least one value is above 2 in accordance with our previous results. As mentioned previously, differences in the estimation approaches used in the different packages can lead to differences in the results, for example, for these fit statistics. In `TAM`, the outfit and infit statistics are conceptually based on Bayesian estimation methods. They use simulations to estimate the expected value of both statistics under the posterior distributions of the person parameters. As a consequence, the outcome may differ slightly when this command is run repeatedly. As in Section 7.3, one could use `set.seed` to make the results reproducible. In addition to the descriptive statistics, `TAM` also reports p-values with and without a Holm correction for the infit and outfit statistics, which agree with the results of the rule of thumb for this example.

`TAM` can also be used for the detection of DIF. We do not present details here, but refer to the TAM online tutorials by Kiefer, Robitzsch, and Wu (2013).

8.3 Person Parameter Estimation

The `person` entry of `tam_rm` contains the expected a posteriori (EAP) estimates (cf. Section 3.5) for the person parameters. We can print the EAP estimates for the first six test takers by entering

```
> head(tam_rm$person)
  pid case pweight score max        EAP    SD.EAP
1   1    1       1     6  14 -0.2176799 0.5505351
2   2    2       1     4  14 -0.8274507 0.5556471
3   3    3       1     4  14 -0.8274507 0.5556471
4   4    4       1     6  14 -0.2176799 0.5505351
5   5    5       1     8  14  0.3912911 0.5542079
6   6    6       1     5  14 -0.5211867 0.5516983
```

The `pid` column gives the person identifier. Here, this is simply the row number in the data matrix. The mean of the posterior distribution, that is the EAP estimator, is presented in column `EAP`, whereas the standard deviation of the posterior distribution is presented in column `SD.EAP`. An alternative function for getting these estimates is the following:

```
> head(IRT.factor.scores(tam_rm))
         EAP    SD.EAP
1 -0.2176799 0.5505351
2 -0.8274507 0.5556471
3 -0.8274507 0.5556471
4 -0.2176799 0.5505351
5  0.3912911 0.5542079
6 -0.5211867 0.5516983
```

We can also get the weighted likelihood (WLE) estimates (cf. again Section 3.5) from the `IRT.factor.scores()` function with `type = "WLE"` (cf. Table 9.1) or by using the `tam.wle()` function by entering

```
> wle_est <- tam.wle(tam_rm)
```

and we can see the first six rows by entering

```
> head(as.data.frame(wle_est))
  pid N.items PersonScores PersonMax      theta     error
1   1      14            6        14 -0.3278524 0.6687089
2   2      14            4        14 -1.2002969 0.6901610
3   3      14            4        14 -1.2002969 0.6901610
4   4      14            6        14 -0.3278524 0.6687089
5   5      14            8        14  0.5645783 0.6829504
6   6      14            5        14 -0.7604822 0.6729278
    WLE.rel
1 0.6382108
2 0.6382108
3 0.6382108
4 0.6382108
5 0.6382108
6 0.6382108
```

Either way we need to reformat the output of the WLE estimation as a data frame, because otherwise the print method defined in TAM will only list summary information about the estimation rather than the estimates for individual test takers themselves.

In the output above, the PersonScores, PersonMax, and theta columns of the WLE output correspond to the score, max, and EAP columns of the EAP output. Whereas the column SD.EAP in the EAP output reports the standard deviation of the posterior distribution of the person parameter, the column error in the WLE output reports the standard error of the person parameter estimates.

8.4 Exercises

1. Bring the data set SPISA from the psychotree package into the workspace. Get the responses for items 37 through 45. These are the responses to the natural science items. Store them under the name natSci_responses and assign them the column names "I 1" through "I 9". (This has already been done in the exercises for the previous chapters.)
2. (Install and) load the TAM package. Fit a Rasch model to the natSci_responses. Display the item parameters. Which items are the easiest and hardest?
3. Look up the name of the function for plotting the expected ICCs (termed "item response functions" in TAM) in Table 9.1 and apply it to your fitted model.
4. Compute the SRMSR as an example of a global indicator of model fit.
5. Display the first few person parameters, estimated by means of the expected a posteriori approach.

9

R Interface to Stan

CONTENTS

In this chapter, we provide a basic example of Bayesian inference for the Rasch model using Stan (Stan Development Team, 2022) and its R interface `rstan` (Stan Development Team, 2021). We will show how to obtain Bayesian estimates and check the model fit using posterior predictive checks. Several additional IRT models are covered in the demonstration function of `rstan`.[1]

Stan is a general language for performing inference in Bayesian models; it is not limited to IRT models. Both Stan and `rstan` can be installed by entering

```
> install.packages("rstan")
```

A potential source of difficulty when first installing Stan is the fact that using Stan requires a working C++ compiler. Thus, you will need to make sure that you have a working C++ compiler before running any of the exercises in this chapter. As of writing, instructions for doing this can be found at

`https://github.com/stan-dev/rstan/wiki/RStan-Getting-Started`

An alternative framework for Bayesian IRT analyses in R is provided by the `brms` package (Bürkner, 2017). We do not present details on the use of this package here, but an introduction in the context of IRT is provided by Bürkner (2021).

[1] In R, this demonstration can be accessed by loading `rstan` via `library("rstan")` and then using the command `stan_demo()`.

9.1 Stan Models

In Stan, we define how the data depends on the model parameters and how the model parameters depend on each other. We know from Chapter 2 that the Rasch model defines the probability of the data matrix using two types of parameters, the person parameters θ_p and the item parameters β_i. In Section 3.4, we defined a prior distribution for the θ_p and β_i, which introduced three new parameters: σ_θ^2, the variance of the person parameters; μ_β, the mean of the item parameters; and σ_β^2, the variance of the item parameters. The joint prior distribution for all of these parameters was defined as

$$
\begin{aligned}
\theta_p \mid \sigma_\theta^2 &\sim N(0, \sigma_\theta^2) \\
\beta_i \mid \mu_\beta, \sigma_\beta^2 &\sim N(\mu_\beta, \sigma_\beta^2) \\
f(\mu_\beta) &\propto 1 \\
\sigma_\theta^2 \mid \nu_\theta &\sim \text{Inv-}\chi^2(\nu_\theta) \\
\sigma_\beta^2 \mid \nu_\beta &\sim \text{Inv-}\chi^2(\nu_\beta)
\end{aligned}
\tag{9.1}
$$

Figure 9.1 shows one way of defining the Rasch model with the prior distributions from Equation (9.1) in Stan. To follow along with the exercises in this chapter, copy the code in this figure into a file named "rasch.stan" (Stan files use the ".stan" extension).

The Stan software takes a Stan file as input and translates the Stan code into C++ code. This C++ code is then compiled and linked into a program implementing an MCMC sampler model. This is why Stan requires a working C++ compiler. Thankfully, once a compiler has been installed, this process is essentially invisible to the user.

Every Stan model is divided into blocks, each defining a different part of the model. Figure 9.1 shows the three required blocks in a Stan model: the `data` block (Lines 1 - 5), which defines the observed data; the `parameters` block (Lines 6 - 12), which defines the model parameters; and the `model` block (Lines 19 - 27), which defines the model. It also shows one optional block, the `transformed parameters` block (Lines 13 - 18). This defines a program for sampling the posterior distribution of the parameters in the `parameters` block given the data in the `data` block. The `model` block defines the prior distribution and the likelihood. The `transformed parameters` block allows us to separate transformations that are useful for the `model` block.

9.1.1 The `data` Block

As already noted, the `data` block defines the observed data. For the Rasch model, the `data` block defines variables for the number of test takers

```
1  data {
2    int<lower=1> num_person;
3    int<lower=1> num_item;
4    int<lower=0, upper=1> U[num_person, num_item];
5  }
6  parameters {
7    vector[num_person] theta;
8    vector[num_item] beta;
9    real mu_beta;
10   real<lower=0> sigma2_theta;
11   real<lower=0> sigma2_beta;
12 }
13 transformed parameters {
14   real<lower=0, upper=1> prob_solve[num_person, num_item];
15   for (p in 1:num_person)
16     for (i in 1:num_item)
17       prob_solve[p, i] = inv_logit(theta[p] - beta[i]);
18 }
19 model {
20   for (p in 1:num_person)
21     for (i in 1:num_item)
22       U[p, i] ~ bernoulli(prob_solve[p, i]);
23   theta ~ normal(0, sqrt(sigma2_theta));
24   beta ~ normal(mu_beta, sqrt(sigma2_beta));
25   sigma2_theta ~ inv_chi_square(0.5);
26   sigma2_beta ~ inv_chi_square(0.5);
27 }
```

FIGURE 9.1: Code for implementing the Rasch model in Stan with the prior distributions defined by Equation (9.1).

`num_person`, the number of items `num_item` and the data matrix `U`. Each distinct piece of data is defined on a separate line. Each line is ended by a semicolon.

Stan requires users to explicitly give the data type for all data and parameters. The `int` preceding `num_person`, `num_item` and `U` indicates that they can only take integer values. We can further restrict the possible values of the data by specifying lower and upper bounds. For example, the statement

```
int<lower=1> num_person;
```

in Line 2 indicates that `N` is an integer greater than or equal to one.

The fact that U is followed by a square bracket indicates that U is an array. The dimensions of U are indicated by integers between the square brackets. Thus, the statement

```
int<lower=0, upper=1> U[num_person, num_item];
```

in Line 4 indicates that U has num_person rows and num_item columns and that each element of U is either zero or one. Specifying num_person and num_item as variables is good practice, since it allows us to reuse the same sampling program for data matrices of different sizes.

9.1.2 The parameters Block

The parameters block defines the model parameters. These parameters are taken directly from Equation 9.1. The parameter theta corresponds to $\boldsymbol{\theta}$, beta to $\boldsymbol{\beta}$, sigma2_theta to σ_θ^2, mu_beta to μ_β, and sigma2_beta to σ_β^2. Parameters must also be given data types. The real data type corresponds to real numbers. Thus, the statement

```
real mu_beta;
```

in Line 9 indicates that mu_beta can be any number. Like integers, we can constrain the values of real numbers as well. For example, the statement

```
real<lower=0> sigma2_beta;
```

in Line 11 indicates that the parameter sigma2_beta is positive. We constrain sigma2_beta, because it is a variance parameter, which by definition cannot take negative values. Finally, we can define vectors of real numbers. For example, the statement

```
vector[num_person] theta;
```

in Line 7 defines a vector of length num_person named theta.

9.1.3 The transformed parameters Block

The transformed parameters block allows us to define transformations of existing variables. Here we use the transformed parameters block to compute the probabilities that test taker p solves item i. The transformed parameters block is split into two parts. First, we declare any parameters whose values will be computed in the transformed parameters block. This must be done before any values can be computed. In this case, we will declare an array with num_person rows and num_item columns named prob_solve. This array stores

$\Pr(U_{pi} = 1 \mid \theta_p, \beta_i)$ for every subject p and item i. This is indicated in Line 14 by the statement

```
real<lower=0, upper=1> prob_solve[num_person, num_item];
```

We then compute the value of each entry in `prob_solve` using the nested loop

```
for (p in 1:num_person)
  for (i in 1:num_item)
    prob_solve[p, i] = inv_logit(theta[p] - beta[i]);
```

in Lines 15 - 17. In Stan, the logistic function is called `inv_logit()`, referring to the fact that it inverts the logit (see Section 2.2.3). Thus, the nested loop states that for every subject `p` between 1 and `num_person` and every item `i` between 1 and `num_item`, `prob_solve[p, i]` is equal to the value of the logistic function evaluated at `theta[p] - beta[i]`.

9.1.4 The `model` Block

The model block defines how the data depends on the parameters and how the parameters depend on each other. This requires translating the likelihood and the prior distribution in Equation (9.1) into Stan's modeling language. The likelihood is defined by the nested loop

```
for (p in 1:num_person)
  for (i in 1:num_item)
    U[p, i] ~ bernoulli(prob_solve[p, i]);
```

in Lines 20 - 22.

The last line,

```
U[p, i] ~ bernoulli(prob_solve[p, i]);
```

indicates that every element of the data matrix `U` has the Bernoulli distribution. As discussed in Section 2.3.2, the Bernoulli distribution is a distribution for a random variable that can only be zero and one. It has a single parameter, which is typically expressed as the probability that the random variable is 1.

The remainder of the `model` block (Lines 23 - 26) specifies the prior distributions for the model parameters given in Equation (9.1). Line 23,

```
theta ~ normal(0, sqrt(sigma2_theta));
```

corresponds to the prior $\theta_p \sim N(0_p, \sigma^2_\theta)$. We need to take the square root of `sigma2_theta`, because Stan parameterizes the normal distribution in terms of its mean and standard deviation. Line 24,

```
beta ~ normal(mu_beta, sqrt(sigma2_beta));
```

corresponds to the prior $\beta_i \sim N(\mu_\beta, \sigma^2_\beta)$.

The last two lines, Lines 25 - 26, specify the prior distributions on σ^2_θ and σ^2_β. Each line indicates that the parameter follows an inverse-χ^2 distribution with 0.5 degrees of freedom. We do not need to explicitly specify a prior distribution for μ_β, because Stan automatically applies a uniform prior distribution to any parameter whose prior is not specified.

9.2 Sampling the Posterior Using RStan

The `rstan` package contains all of the tools we need to create a sampling program, execute it and analyze the results. We can bring the `rstan` package into the workspace by entering

```
> library("rstan")
```

The `rstan` package contains the function `stan()` which creates a sampling program and executes that program for the data supplied, generating samples from the posterior distribution of the model.

We will use Stan to sample the posterior distribution of the model parameters (cf. Section 3.4) given the responses of the first 400 test takers from the FIMS data. This is the same sample that we fit using conditional maximum likelihood in Chapter 6 and marginal maximum likelihood in Chapter 7 and Chapter 8. We can obtain the responses for the first 400 test takers by entering

```
> data(data.fims.Aus.Jpn.scored, package = "TAM")
> people <- 1 : 400
> responses <- data.fims.Aus.Jpn.scored[people, 2:15]
> colnames(responses) <- gsub("M1PTI", "I", colnames(responses))
```

To run a Stan program, we need to provide all of the arguments defined in the `data` block. This means we need to provide the number of test takers `num_person`, the number of items `num_item` and the data matrix `U`. These arguments are passed to Stan as a list. We can create the list for the FIMS data by entering

```
> fims_args <- list(num_person = nrow(responses),
+                   num_item = ncol(responses),
+                   U = responses)
```

We can sample the posterior of the Rasch model by entering

```
> stan_rm <- stan("rasch.stan", data = fims_args)
```

Note that this line of code will only work when R's working directory is the directory containing the Stan file "rasch.stan". You can get and set R's working directory using the `getwd()` and `setwd()` functions. If you are using RStudio, you can also open the Stan file "rasch.stan" and then choose "Set Working Directory ▶ To Source File Location" from the "Session" menu. Also note that all commands in this section will take some time to run.

Though MCMC techniques eventually produce samples whose distribution is representative of the posterior, the initial MCMC samples are typically not. In practice, we deal with this by throwing away the first B samples. These are often called "burn-in" or "warm-up" samples. The number of warm-up samples needed depends on a number of factors, including correlation between model parameters in the joint posterior, the initial values chosen for the model parameters, the MCMC algorithm used and whether that algorithm uses an initial "adaptation" period to tune parameters and increase efficiency.

When first working with a model, a good rule of thumb is to use the first half of the samples as a warm-up. The number of warm-up samples can be set using the `warmup` argument to the `stan()` function. It defaults to half the number of iterations. The number of iterations is set with the `iter` argument. By default, this argument is set to 2000. We left both of these parameters at their default values when fitting the FIMS data, thereby using 1000 warm-up samples.

In addition to using a warm-up sample, we also use convergence diagnostics to check whether the samples we've drawn appear representative of the posterior distribution. The most common convergence diagnostic is the $\hat{R}$ statistic (Gelman, Stern, Dunson, Vehtari, & Rubin, 2013). To compute the $\hat{R}$ statistic, we split the desired number of samples among 3 – 5 "chains". Each chain is seeded with a different starting value – typically, an independent sample from the prior distribution – and sampled independently. As a result, the samples contained in one MCMC chain will be independent of the samples contained in every other chain. `stan()` uses four MCMC chains by default, but this number can be changed using the `chains` argument.

Suppose we start the MCMC chains in disparate regions of the posterior distribution. MCMC techniques are iterative, so the sample drawn on one iteration depends on what was sampled on the previous iteration. This dependence, called *autocorrelation*, can persist over many samples. MCMC chains with high autocorrelation slowly drift about the parameter space, eventually converging on the posterior distribution. As a result, when we start the chains far apart from each other, the initial distributions of the chains' samples will look very different from one another. After we run each chain for a large number of iterations, the chains will have "forgotten" their starting values, and the distributions of the chains' samples will look similar to one another, because, essentially, they are all sampling from the same distribution, the posterior.

The $\hat{R}$ statistic measures the degree to which the chains have forgotten their initial values, a property known as "mixing". It is defined as the square root of the ratio between the variance of the combined MCMC sample, $\widehat{\text{Var}}_{\texttt{total}}$, and the average variance within each MCMC chain, $\widehat{\text{Var}}_{\texttt{within}}$. When the chains are far apart, the variance of the combined sample will be much larger than variance within the chains, because the variance of the combined sample will also contain the variance among the locations of the chains. In this case, $\hat{R}$ will be much larger than one. When the chains have mixed, their locations will be essentially the same, so $\widehat{\text{Var}}_{\texttt{total}} \approx \widehat{\text{Var}}_{\texttt{within}}$ and $\hat{R} \approx 1$. In practice, a cutoff of 1.1 is used to assess convergence. When $\hat{R} < 1.1$ for all of our model parameters, we have no reason to believe that our samples are not representative of the posterior distribution. If some of the $\hat{R}$ values are larger than 1.1, increasing the number of iterations will often fix the problem. See a textbook on Bayesian inference, such as Kruschke (2014), for a more general discussion of mixing and MCMC diagnostics.

For models with a small number of parameters, we can easily check $\hat{R}$ values by printing the object returned by `stan()`, e.g.,

```
> stan_rm
```

In our case, the output of this would be very long, since there are a large number of model parameters. We can check that $\hat{R} < 1.1$ for all of the model parameters programmatically using the `summary()` function. Calling `summary()` on the output of a call to `stan()` computes $\hat{R}$ values for every model parameter and stores it in the "Rhat" column of the "summary" element in the returned object. To check whether $\hat{R} < 1.1$ for all of the model parameters, we define a custom function as follows.

```
> check_rhats <- function(stan_fit) {
+   stats <- summary(stan_fit)
+   all(stats$summary[, "Rhat"] < 1.1)
+ }
>
> check_rhats(stan_rm)
[1] TRUE
```

This functionality is implemented here as a function to make it easier to reuse for the exercises. The function returns `TRUE` when all of the $\hat{R}$ values are below 1.1 and `FALSE` otherwise. This returns `TRUE` for our fit to the FIMS data, indicating the MCMC chains have mixed.

We can view the item parameter estimates by printing **stan_rm**. Explicitly calling **print()** allows us to restrict printing to specific model parameters. We can print information about just the item parameters by entering

```
> print(stan_rm, pars = "beta")
Inference for Stan model: rasch.
4 chains, each with iter=2000; warmup=1000; thin=1;
post-warmup draws per chain=1000, total post-warmup draws=4000.

          mean se_mean   sd  2.5%   25%   50%   75% 97.5% n_eff Rhat
beta[1]  -1.10       0 0.13 -1.34 -1.18 -1.10 -1.01 -0.85  5223    1
beta[2]  -1.24       0 0.13 -1.50 -1.34 -1.24 -1.15 -0.99  5481    1
beta[3]  -2.03       0 0.16 -2.34 -2.13 -2.03 -1.92 -1.72  5715    1
beta[4]  -0.05       0 0.12 -0.28 -0.13 -0.05  0.03  0.19  5350    1
beta[5]   2.51       0 0.19  2.16  2.39  2.51  2.64  2.89  6713    1
beta[6]  -1.24       0 0.13 -1.50 -1.33 -1.25 -1.15 -0.99  4797    1
beta[7]   0.81       0 0.12  0.56  0.73  0.81  0.89  1.05  5072    1
beta[8]  -0.50       0 0.12 -0.72 -0.57 -0.50 -0.41 -0.26  5234    1
beta[9]   1.32       0 0.13  1.07  1.23  1.32  1.41  1.59  5010    1
beta[10] -0.40       0 0.12 -0.63 -0.48 -0.40 -0.32 -0.15  5646    1
beta[11]  2.05       0 0.16  1.75  1.94  2.05  2.16  2.39  5848    1
beta[12]  1.75       0 0.15  1.47  1.65  1.75  1.84  2.05  6323    1
beta[13]  2.40       0 0.17  2.06  2.28  2.39  2.52  2.74  6744    1
beta[14] -1.93       0 0.16 -2.24 -2.03 -1.92 -1.82 -1.62  5811    1
```

The first few rows of the output are about the model in sampling process. The first line tells us the name of the Stan model. Because we saved our model in a file named "rasch.stan", Stan automatically named our model "rasch". The next line tells us the number of chains (4), the total number of iterations per chain (2000), the number of warm-up samples (1000) and the thinning interval (1). When there is substantial autocorrelation between successive MCMC samples, we need a large number of samples to get a complete view of the posterior distribution. To simplify computation, people often only retain every t-th sample. The number t is known as the thinning interval. For example, a thinning interval of 5 means we keep every 5th sample and discard the remainder.

The third line lists the total post warm-up draws per chain (1000) and in total (4000). The total number of post warm-up draws per chain is equal to the total number of iterations per chain minus the number of warm-up samples divided by the thinning interval. The total number of post warm-up draws is equal to the total number of post warm-up draws per chain times the number of chains.

The remaining output displays sample statistics for each item parameter in the 4000 MCMC samples. Each row of the output corresponds to an item parameter, indicated by the row name to the left. The columns give the values of the sample statistics. The first column gives the sample mean, the second its standard error, and the third the sample standard deviation. The next five columns give the 2.5, 25, 50, 75, and 97.5 percentiles of the MCMC samples. The last two columns list the effective sample size and the (rounded) $\hat{R}$.

As discussed, the MCMC samples within a chain can be correlated. The effective sample size, `n_eff`, can roughly be thought of as the equivalent number of independent samples. Thus, when `n_eff` is small, the mean and other posterior statistics are going to be noisy estimates of their true value. This means that we should not interpret parameter values when `n_eff` is small. Instead, you should increase the number of samples or take steps to reduce the correlation between samples.

Here we see that the effective sample sizes can also exceed the number of MCMC samples, which happens when successive MCMC samples are negatively correlated. We only need to take remedial measures when effective sample sizes are small. In this example, all of the effective sample sizes are large enough for the sample statistics to be reasonable estimates of mean, standard deviation, and 2.5, 25, 50, 75, and 97.5 percentiles of the marginal posteriors for the item parameters.

These statistics provide us with a lot of information about the marginal posterior distributions of the item parameters. The posterior means are a point estimate of the item parameters, similar to the estimates we've obtained using other methods. Looking at the `mean` column, we see that the posterior mean of β_1 is -1.1, the posterior mean of β_2 is -1.24, etc. These estimates agree with our estimates from other chapters. Item 3 has the lowest difficulty and item 7 (given by `beta[5]`, as the FIMS items are not numbered consecutively and item 7 is the fifth of the items) has the highest difficulty.

We can read HPD intervals from the "2.5%" and "97.5%" columns. These columns indicate that a 95% HPD interval for β_1 ranges from -1.34 to -0.85. This means that there is a 95% chance that the difficulty of item 1 is between -1.34 and -0.85 given the observed responses. Finally, we can read the median from the "50%" column (recall that the median is the 50th percentile). Thus, the median of β_1 is also -1.1, suggesting the distribution is symmetric. In fact, by comparing the sample quantiles of β_1 to the quantiles of a normal distribution with a mean of -1.1 via

```
> targ_qtls <- c(0.025, .25, .5, .75, 0.975)
> qnorm(targ_qtls, -1.1, .13)
[1] -1.3547953 -1.1876837 -1.1000000 -1.0123163 -0.8452047
```

we see that the marginal posterior distribution of β_1 is approximately normally-distributed with a mean of -1.1 and a standard deviation of 0.13.

Bayesian methods estimate the person and item parameters simultaneously. We can view the same summary for the person parameters, but the output will be much longer. We will not show it, but you can view the output yourself by entering

```
> print(stan_rm, pars = "theta")
```

Given the large number of ability parameters, it may be more interesting to view summary statistics of posterior means of the ability parameters to understand the range of these estimates. We can get these estimates using the `summary()` function. For example, we can summarize of the posterior means of the person parameters by entering

```
> mcmc_stats <- summary(stan_rm)$summary
> summary(mcmc_stats[startsWith(rownames(mcmc_stats), "theta"),
+         "mean"])
      Min.    1st Qu.     Median       Mean    3rd Qu.
-1.8101317 -0.5222705  0.0825681 -0.0002951  0.3994391
      Max.
 2.3865552
```

Here the mean estimates are stored in the "summary" element of the result of `summary(stan_rm)`. This element is a matrix with one row per parameter and one column per summary statistic (these are the same summary statistics displayed by the print method). Finally, the `startsWith()` function returns `TRUE` whenever its first argument (here `rownames(mcmc_stats)`) starts with the string given to the second argument (here `"theta"`). Thus, `mcmc_stats[startsWith(rownames(mcmc_stats), "theta"), "mean"]` returns a vector of posterior means of the person parameters. Passing this vector to the `summary()` function computes the minimum, first quartile, median, mean, third quartile and maximum of its values. Thus, the estimated person parameter values range from −1.8 to 2.4, with a mean value of effectively zero.

9.3 Evaluating Goodness-of-Fit

In this section, we demonstrate how we can compute and use the posterior predictive to assess goodness-of-fit when using Bayesian inference for the Rasch model (cf. Section 4.3.8), using point biserial correlation as an example. For a given item, the point biserial correlation is defined to be the correlation between the observed scores and its pattern of responses. Sinharay et al. (2006) demonstrated that the point biserial correlation was effective in evaluating the fit of the Rasch model, as compared to more complex models, such as the 2PL. We also suggest reading Sinharay (2005) and Sinharay et al. (2006) for additional methods of evaluating the Rasch and other IRT models using the posterior predictive distribution.

We can compute the point biserial correlation using the following function:

```
> biserial_cor <- function(U) {
+   as.vector(cor(U, rowSums(U)))
+ }
```

In this function, the `rowSums(U)` computes the observed scores. These are passed along with `U` to the `cor()` function to compute the correlation between the observed scores and the response patterns for each item. In this case, `cor(U, rowSums(U))` returns a 14-by-1 matrix. We would rather return a vector, so we pass the result to `as.vector` to drop the second dimension.

To compute the observed biserial correlation, we simply need to pass the responses to the `biserial_cor()` function as follows (here we are also using parentheses to print the results in the same step).

```
> (obs_cor <- biserial_cor(responses))
 [1] 0.49425385 0.55709696 0.41392738 0.57149367 0.36155602
 [6] 0.52945776 0.43783925 0.44141406 0.39495680 0.44143341
[11] 0.49759495 0.09217772 0.18948893 0.44580662
```

These values are shown by the closed circles in Figure 9.2 on page 188. We see most of the observed biserial correlations are around 0.4. However, we can also see that the biserial correlations corresponding to the 12th and 13th items are considerably lower, respectively, 0.09 and 0.19.

The point biserial correlation is closely related to an item's discrimination. Small discrimination values lead to point biserial correlations near zero, whereas large discrimination values lead to correlations near one. In the Rasch model, every item has the same discrimination by definition. Thus, we would expect to see similar values for the point biserial correlation across items for data generated by the Rasch model. The two FIMS items with small point biserial correlation values could be cause for alarm.

We can check by comparing the observed point biserial correlations with their posterior predictive distribution. In Figure 9.2, for example, open circles show the median of the posterior predictive distribution for that item's point biserial correlation and the dashed lines show a 90% HPD interval. We can see that the 12th and 13th items (corresponding to FIMS items 21 and 22) fall below their respective 90% HPD intervals, indicating a lower discrimination. A few other items also have observed point biserial correlations that are more extreme than we would have expected under the Rasch model, including most prominently the 2nd, 4th, and 11th item (corresponding to FIMS items 2, 6, and 19), which fall above their respective 90% HPD intervals, indicating a higher discrimination. These results are in line with our previous findings. For example, the empirical ICCs in Section 6.2.2 showed that the 12th item (item 21) had a flatter empirical ICC, and that the 11th item (item 19) had a steeper empirical ICC than expected under the Rasch model. This suggests that the Rasch model cannot provide a good account of the observed point biserial correlations and that a model with a free discrimination parameter, such as the 2PL or 3PL, would do a better job.

We can compute the posterior predictive distribution of each item's point biserial correlation by applying a two-step process to each posterior sample, which in the following we will denote by k. First, we use the $P \times I$ matrix of solution probabilities to simulate a response matrix $\tilde{\boldsymbol{U}}^{(k)}$. Then, we compute each item's point biserial correlation in $\tilde{\boldsymbol{U}}^{(k)}$. Applying this process to every posterior sample samples the posterior predictive distribution of the point biserial correlation.

How does this relate to our definition from Section 4.3.8? Recall that

$$\Pr(\tilde{\boldsymbol{U}} \mid \boldsymbol{U}) = \int_{\boldsymbol{\theta}} \int_{\boldsymbol{\beta}} \Pr(\tilde{\boldsymbol{U}} \mid \boldsymbol{\theta}, \boldsymbol{\beta}) \cdot f(\boldsymbol{\theta}, \boldsymbol{\beta} \mid \boldsymbol{U}) d\boldsymbol{\beta} d\boldsymbol{\theta},$$

where $f(\boldsymbol{\theta}, \boldsymbol{\beta} \mid \boldsymbol{U})$ is the joint posterior distribution of $\boldsymbol{\theta}$ and $\boldsymbol{\beta}$ given the observed response matrix $\boldsymbol{U}$ and $\Pr(\tilde{\boldsymbol{U}} \mid \boldsymbol{\theta}, \boldsymbol{\beta})$ is the probability of an arbitrary response matrix $\tilde{\boldsymbol{U}}$ under the Rasch model. We have just drawn samples from $f(\boldsymbol{\theta}, \boldsymbol{\beta} \mid \boldsymbol{U})$ using Stan. We can now sample the posterior of $\tilde{\boldsymbol{U}}$ by sampling a new response matrix for each of these posterior samples, thereby marginalizing over $\boldsymbol{\theta}$ and $\boldsymbol{\beta}$. We can sample the posterior predictive distribution of a summary statistic like the point biserial correlation by first drawing a response matrix $\tilde{U}$ according to $\Pr(\tilde{\boldsymbol{U}} \mid \boldsymbol{U})$ and then computing the summary statistic for that response matrix.

We can implement this in R as follows. First, we extract the solution probabilities from the Stan model fit using `rstan`'s `extract()` function.

```
> prob_solve <- extract(stan_rm, "prob_solve")[[1]]
```

We have added the `[[1]]` at the end, because the `extract()` function returns a list. We can get the first element of that list by suffixing the call to `extract()` with `[[1]]`. We can use the `str()` function, i.e.,

```
> str(prob_solve)
 num [1:4000, 1:400, 1:14] 0.607 0.585 0.731 0.635 0.849 ...
 - attr(*, "dimnames")=List of 3
  ..$ iterations: NULL
  ..$           : NULL
  ..$           : NULL
```

to see that `prob_solve` is a 4000-by-400-by-14 array. This tells us that the first dimension corresponds to iteration and that, for each iteration, we have a single 400-by-14 array of probabilities. Thus, `prob_samp[k, p, i]` is the kth sample of the probability that test taker p correctly responds to item i. As with the summary statistics in the previous section, the elements of this matrix will differ from run to run.

Second, we define a function to sample a random matrix of responses given a matrix of solution probabilities.

```
> sample_responses <- function(soln_pr) {
+   array(rbinom(length(soln_pr), 1, soln_pr), dim(soln_pr))
+ }
```

The input to this function is a matrix of solution probabilities for the items of a test. The output is a random matrix of item responses having the same dimension as `soln_pr` and response probabilities given by the elements of `soln_pr`. These samples are generated as follows. First, we sample a vector of responses of length `length(soln_pr)` via `rbinom(length(soln_pr), 1, soln_pr)`. The first P elements of this vector give the responses of the P test takers to the first item, the second P elements give the responses of the test takers to the second item, etc. The next step reshapes the result to have the same size as `soln_pr` using the `array()` function.

We can sample a new response matrix for each sample of `prob_solve` using a `for` loop. On each iteration of the loop, we generate a new point biserial correlation sample from its posterior predictive distribution. The R code is shown below.

```
> pred_cor <- array(NA, dim(prob_solve)[c(1,3)])
> for (k in 1 : nrow(prob_solve)) {
+   U_pp <- sample_responses(prob_solve[k,,])
+   pred_cor[k,] <- biserial_cor(U_pp)
+ }
```

The first line of code creates a matrix `pred_cor` that will be used to store the samples from the posterior predictive distribution of point biserial correlations. Each element `pred_cor[k, i]` contains the value of the point biserial correlation of item i for the response matrix sampled from the kth posterior sample. The elements of `pred_cor` are initially set to `NA`.

The second line defines the `for` loop. A `for` loop repeatedly runs the code between the curly braces. As defined, R will create a variable `k` that is only available for code between the curly braces. The first time R runs the code between the braces, the value of `k` is 1. The second time, it is 2, the third time 3, etc., continuing until `k` is equal to the number of rows in `prob_solve`. After it finishes running the code with `k` equal to `nrow(prob_solve)`, the number of posterior samples, R exits the curly braces and stops.

The third and fourth lines list the tasks that are completed for each value of `k`. Here each value of `k` corresponds to the index of a posterior sample. Thus, the first time R runs the third and fourth lines of code, it starts by generating the first sample from the posterior predictive distribution over response matrices by calling `sample_responses()` for the response probabilities from our first MCMC sample, `prob_solve[1,,]`, and then saves the result into the variable `U_pp`. Next, it computes the point biserial correlation for the sampled response matrix `U_pp` and stores the result in the first row of `pred_cor`. This process is repeated for the second MCMC sample, the third, etc., until we've sampled a point biserial correlations for each MCMC sample and stored it in the corresponding row of `pred_cor`.

These steps are encapsulated into the following function, provided to make it easier to use the point biserial correlation for evaluating the fit of the Rasch model.

```
> sample_biserial_ppd <- function(stan_fit) {
+   prob_solve <- extract(stan_fit, "prob_solve")[[1]]
+   pred_cor <- array(NA, dim(prob_solve)[c(1,3)])
+   for (k in 1 : nrow(prob_solve)) {
+     U_pp <- sample_responses(prob_solve[k,,])
+     pred_cor[k,] <- biserial_cor(U_pp)
+   }
+   pred_cor
+ }
```

To use the function, simply pass the object returned from fitting the Rasch model using the Stan code provided. We could compute the posterior predictive distribution for our example by passing **stan_rm** as follows:

```
> pred_cor <- sample_biserial_ppd(stan_rm)
```

The code used to plot the observed and predicted point biserial correlations in Figure 9.2 is provided in the function below.

```
> plot_post_pred_items <- function(obs_cor, pred_cor) {
+   n_item <- ncol(pred_cor)
+   # Compute the median and 90% interval for posterior predictive
+   # point biserial correlations
+   int_pr <- c(.05, .95)
+   len_int_pr <- length(int_pr)
+   meds <- numeric(n_item)
+   pp_ints <- matrix(NA, n_item, len_int_pr)
+   for (i in 1:n_item) {
+     meds[i] <- median(pred_cor[,i])
+     pp_ints[i,] <- quantile(pred_cor[,i], int_pr)
+   }
+   # Plot medians, intervals, observed correlations
+   plot(meds, xlab = "Item", ylab = "Point Biserial Correlation",
+        xaxt = "n", ylim = c(0, 1))
+   axis(side = 1, at = 1:14)
+   for (i in 1:n_item) {
+     lines(rep(i, len_int_pr), pp_ints[i,], lty = 3)
+   }
+   points(obs_cor, pch = 19)
+ }
```

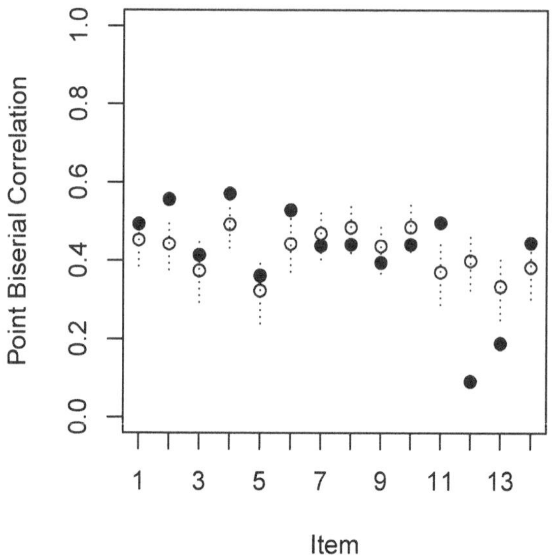

FIGURE 9.2: Comparison of observed and predicted point biserial correlations. Each closed circle shows the point biserial correlation for a single test item. Each open circle shows the median of the posterior predictive distribution of the point biserial correlations, and the dashed vertical lines show a 90% HPD interval.

This function takes the observed point biserial correlations to its `obs_cor` argument and a matrix of samples from its posterior predictive distribution to its `pred_cor` argument. Thus, before calling this function, we need to have already computed both values. We could recreate the plot from our example as follows.

```
> obs_cor <- biserial_cor(responses)
> pred_cor <- sample_biserial_ppd(stan_rm)
> plot_post_pred_items(obs_cor, pred_cor)
```

The function `plot_post_pred_items()` works as follows. It starts by computing the median and 5th and 95th percentiles of each item's posterior predictive distribution. Then, it plots the medians as open circles using the `plot()` function. The `xlab` and `ylab` arguments set the x- and y-axis labels to "Item" and "Biserial", respectively. The `ylim` argument sets the limits of the y-axis to be zero and one. The `xaxt` argument tells `plot()` not to draw the x-axis. This axis is added manually by the call to `axis()` on the next line.

Next we plot the dashed vertical lines using the `lines()` function. The first two arguments to this function give the x and y coordinates of the points we would like connected via a line. The `lty` argument sets the line type in the plot to 3, corresponding to the dashed lines shown. Finally, we add the observed point biserial correlations to the plot as closed circles using the `points()` function. The `pch` argument sets the plot marker style to 19, corresponding to a closed circle.

9.4 Exercises

1. Bring the data set `SPISA` from the `psychotree` package into the workspace. Get the responses for items 37 through 45. These are the responses to the natural science items. Store them under the name `natSci_responses` and assign them the column names "I1" through "I9". (This has already been done in the exercises for the previous chapters.)
2. (Install and) load the `rstan` package. As outlined in the beginning of this chapter, you may need to install a C++ compiler. Copy the code in Figure 9.1 to a file named `rasch.stan`. Switch R's working directory to the directory containing this file.
3. Use the `stan()` function, together with the commands in Figure 9.1, and specify the `data` argument to fit a Rasch model to the SPISA data. *Hint:* You may receive a warning about the bulk effective sample size being low. If you receive this warning, increase the number of MCMC iterations to 4000.
4. Use the $\hat{R}$ statistics to determine if the samples are representative of the posterior distribution.
5. Print the output for the item parameters. Which item is the easiest? Which item is the hardest? Why?
6. What is the 95% HPD interval for the difficulty of item 3? What is the 50% HPD interval?
7. Sample the posterior predictive distribution of the point biserial correlation for the item parameters and plot each item's observed point biserial correlation against the median and a 95% HPD interval for its value, as in Section 9.3. Do you think the Rasch model provides a reasonable account of the SPISA data?

Summary of R Commands for eRm, mirt, *and* TAM

The table on the following pages lists the R commands for the most common operations in Rasch analyses for the three packages `eRm`, `mirt`, and `TAM`. (Stan commands are not included here explicitly because their structure is different from classical R commands.)

This table is not an exhaustive list of functions for either package, and, since R packages are continuously updated by their authors, it can only serve as a snapshot of what the packages offer today, and may be outdated tomorrow. For more detailed and more up-to-date information, please consult the packages' help pages.

Besides the R packages for Rasch/IRT analyses used in this book, there is a variety of additional packages for other psychometric analyses. A regularly updated list can be found on the CRAN Task View website on Psychometric Models and Methods:

`https://cran.r-project.org/web/views/Psychometrics.html`

	eRm	mirt	TAM	other
Item parameter estimation				
CML[1]	`RM()`			
MML[2]		`mirt(...,`	`tam.mml()`	
		`itemtype = "Rasch")`		
Person parameter estimation				
ML	`person.parameter()`	`fscores(...,`	`IRT.factor.scores(...,` [3]	
		`method = "ML")`	`type = "MLE")`	
WLE		`method = "WLE")`	`type = "WLE")`	
EAP		`method = "EAP")`	`type = "EAP")`	
MAP		`method = "MAP")`		
Joint item and person parameter estimation				
JML			`tam.jml()`[4]	
Bayesian				Stan

[1] Applicable also for other Rasch-type models. For example, for PCM use function `PCM()` in `eRm`.
[2] Applicable also for non-Rasch models. For example, for 2PL use functions `mirt(..., itemtype = "2PL")` or `tam.mml.2pl()`.
[3] `TAM` offers several different functions for person parameter estimation, but `IRT.factor.scores` includes all approaches in one function.
[4] JML estimation in `TAM` is currently only available for the Rasch model.

	eRm	mirt	TAM	other
Graphical assessment				
Person item map	`plotPImap()`			
ICCs (single items)	`plotICC()`	`plot(..., type = "trace")`	`IRT.irfprobPlot()`	
ICCs (all items)	`plotjointICC()`	`itemplot()`		
Empirical ICCs	`plotICC(..., empICC = list("raw"))`	`empirical_plot()`		
Graphical test	`plotGOF()`			
Item/test information	`plotINFO()`	`iteminfo()`	`IRT.informationCurves()`	
Tests for item and person invariance				
Andersen's LR test	`LRtest()`			
Martin-Löf test	`MLoef()`			
Wald Test	`Waldtest()`	`DIF()`		`psychotools`
Anchor selection				`psychotools`
Further DIF tests	`stepwiseIt()`	`DIF()`		`difR`

	eRm	mirt	TAM	other
Goodness-of-fit tests and statistics				
Item and item pair fit		`residuals()`	`IRT.modelfit()`	
	`itemfit()`	`itemfit()`	`tam.fit()`	
	itemwise χ^2; in/outfit	various itemwise statistics, in/outfit (via argument `fit_stats`)	various Q_3 based statistics for item pairs; in/outfit	
M_2		`M2()`		
Person fit	`personfit()`	`personfit()`	`tam.personfit()`	
	χ^2, in/outfit	z_h, in/outfit	in/outfit	
Nonparametric tests	`NPtest()`			
Posterior predictive checks				Stan
Separation indices				
Person separation index G_P	`SepRel()`			
Model comparisons				
Likelihood ratio test		`anova()`	`anova()`	
Information criteria	`IC()`	`anova()`	`anova()`	
			`IRT.modelfit()$IRT.IC`	

TABLE 9.1: Summary of most common R commands for `eRm`, `mirt`, and `TAM`.

Part III

Beyond the Rasch Model

10

Extensions to the Rasch Model

CONTENTS

In the remainder of this book, we will cover a variety of extensions to the Rasch model, which serve different purposes. We will not go into as much detail in this part as we did in the first part of the book, because our intention here is merely to give an overview. However, on each topic we provide references where the interested reader can find additional information. In this first chapter of Part III, we will present a variety of conceptual extensions to the Rasch model for dichotomous items. However, many of their principles generalize to other IRT models and/or other response formats.

For example, the linear-logistic test model presented in Section 10.1 allows for different cognitive operations that are required to solve an item. Section 10.2 introduces models for describing differences between respondents, and Section 10.3 discusses models for items that aim to assess more than a single latent trait.

One note on notation before we proceed: In the following, we will be reusing some of the Greek and Latin letters we have used in the first part of the book with a different meaning – and accordingly with different indices. We are aware that this is not fully consistent, but it will make it easier to read the following parts and relate them to common notation in the literature.

10.1 The Linear-Logistic Test Model

Fischer's (1973) linear-logistic test model (LLTM) describes each item's difficulty parameter as a linear combination of certain characteristics. These characteristics could be, e.g., cognitive operations that are required for solving an item. A test of mathematical ability, for example, may require mastery of addition, multiplication, and operator precedence. Solving each test item requires mastery of one or more of these cognitive operations.

Item	Problem	Addition	Multiplication	Precedence
1	3×2	0	1	0
2	6×2	0	1	0
3	$8 + 3$	1	0	0
4	$3 \times 2 + 4$	1	1	1
5	$3 + 2$	1	0	0

TABLE 10.1: Cognitive operations required to solve each of five items.

Table 10.1 shows the cognitive operations required to solve five items of a mathematics test comprising the type of problems given in the "Problem" column. The "Addition", "Multiplication", and "Precedence" columns indicate whether the addition, multiplication, and operator precedence operations are required to solve the item. For example, Item 1 (3×2) requires multiplication, but not addition or operator precedence. Thus, the "Multiplication" column is 1 for Item 1 and the remaining columns are 0. Alternatively, Item 4 ($3 \times 2 + 4$) requires addition, multiplication, and operator precedence. Thus, all three columns have ones.

In the LLTM, each required cognitive operation a increases the difficulty by an amount η_a. Our example test includes three cognitive operations. Thus, we have three different values of η_a: η_1, the increase associated with addition, η_2, the increase associated with multiplication, and η_3, the increase associated with operator precedence. The difficulty of an item is the sum of the η_a values associated with the cognitive operations it requires. This means that the difficulty of items 1 and 2 is η_2, the difficulty of items 3 and 5 is η_1, and the difficulty of item 4 is $\eta_1 + \eta_2 + \eta_3$.

More generally, each entry in Table 10.1 can be represented as a weight w_{ia}. The value of w_{ia} is one when item i requires cognitive operation a and is zero when it does not. The difficulty of an item is the sum

$$\beta_i = \sum_{a=1}^{A} w_{ia} \cdot \eta_a.$$

Here, A is the total number of cognitive operations required to solve the test items. For our example, $A = 3$.

We can verify this equation for the items in Table 10.1. For items 1 and 2, we have $w_{i1} = 0$, $w_{i2} = 1$, and $w_{i3} = 0$, so

$$\beta_i = \sum_{a=1}^{A} w_{ia} \cdot \eta_a = 0 \cdot \eta_1 + 1 \cdot \eta_2 + 0 \cdot \eta_3 = \eta_2.$$

For items 3 and 5, we have $w_{i1} = 1$ and $w_{i2} = w_{i3} = 0$, so

$$\beta_i = \sum_{a=1}^{A} w_{ia} \cdot \eta_a = 1 \cdot \eta_1 + 0 \cdot \eta_2 + 0 \cdot \eta_3 = \eta_1.$$

Finally, for Item 4, we have $w_{i1} = w_{i2} = w_{i3} = 1$, so

$$\beta_i = \sum_{a=1}^{A} w_{ia} \cdot \eta_a = 1 \cdot \eta_1 + 1 \cdot \eta_2 + 1 \cdot \eta_3 = \eta_1 + \eta_2 + \eta_3.$$

The operation difficulties η_a in the LLTM can be estimated in the same way as the difficulties in the Rasch model. In R, the LLTM is available in the `eRm` package. The weights w_{ia} are typically specified by content experts when the items are constructed. The combination of weights and operation difficulties can be used to better understand test taker performance. It can also be used to automatically generate items with known difficulties (e.g., Holling, Bertling, & Zeuch, 2009).

10.2 Modeling Differences Between People

Another class of Rasch model extensions deals with the case where there are multiple groups of test takers, each having different item parameters. The following approaches to deal with these multiple groups can be considered as more general ways of testing for DIF than those introduced in Section 4.2. While there the groups of test takers were specified a priori, the following methods are more flexible. They can detect groups that are latent or formed by combinations of more than one covariate.

10.2.1 The Mixture Rasch Model

An important approach for accounting for differences between groups of test takers is the mixture[1] Rasch model (Rost, 1990). It is used for finding latent groups of people sharing the same model parameters.

[1]The mixture Rasch model is sometimes also referred to as the "mixed" Rasch model because this name was used in the original literature. However, the term mixed models is more commonly used for models with random effects, which we will treat in Section 10.2.3. The mixture Rasch model, by contrast, is a mixture distribution model.

The model assumes that there are G latent groups (or latent classes) of people and that, within each group, test-taking behavior can be accounted for by the Rasch model. Typically, we do not know the number of latent classes in advance. We determine the number of classes by applying the mixture Rasch model for different values of G and using information criteria to select the clustering providing the best account of the observed data. When the number of latent classes found by this procedure is greater than one, and ability differences are appropriately accounted for (Frick, Strobl, & Zeileis, 2015), this indicates DIF between the latent classes.

After finding latent classes, we often want to know what the people in each class have in common. A second step in the analysis could then be to try to describe the latent classes based on observable covariates such as age and gender (e.g., Cohen & Bolt, 2005). Alternatively, we could model the probability of latent-class membership using explanatory variables (Dayton & Macready, 1988). These variables are often called concomitant variables, because they tend to occur along with class membership.

The mixture approach can also be used to detect different response strategies. For example, the responses to a personality or attitude test with polytomous items may contain one group of people who tend to offer extreme responses, such as "strongly agree" or "strongly disagree", and another group who tend not to give extreme responses (Kaplan & Saccuzzo, 2013). These different groups could be identified with the help of a mixture model.

Mixture Rasch models can be estimated using the R packages `mixRasch` (Willse, 2014) and `psychomix` (Frick, Strobl, Leisch, & Zeileis, 2012). The psychomix package is particularly flexible, allowing the user to include concomitant variables to describe the latent classes.

10.2.2 Model-Based Recursive Partitioning

Model-based recursive partitioning is another approach for detecting groups of test takers with different item parameters. This approach is also known as Rasch trees. In contrast to the mixture Rasch model, the groups found by this method are not latent, but are based on combinations of observed covariates, such as age and gender (Strobl, Kopf, & Zeileis, 2015). Strobl, Malley, and Tutz (2009) give a general introduction to recursive partitioning methods.

Model-based recursive partitioning operates by recursively splitting the test takers into groups, based on statistical tests for parameter instability. For example, if a test finds a difference between the item parameters of younger and older test takers, the sample will be split at the age corresponding to the strongest parameter difference. The resulting groups are then tested for differences in other covariates. The final result is a tree structure, which can be represented graphically and interpreted in a straightforward way.

This is illustrated by means of a small artificial example in Figure 10.1. Here we see that some item parameters in a test of 20 items significantly differ between three groups of test takers: males up to the age of 34, males over the

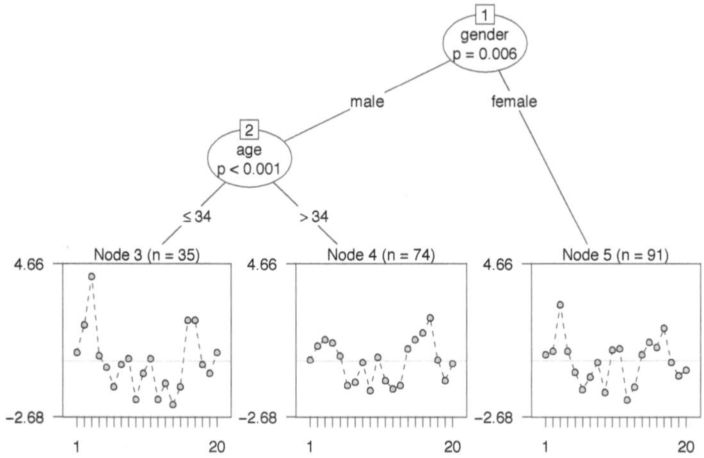

FIGURE 10.1: Example output for model-based recursive partitioning.

age of 34, and females. The upper part of Figure 10.1 displays how the test takers are recursively split into groups based on significant item parameter differences. In each node, the covariate used for splitting is displayed together with the p-value of the underlying test for parameter instability. The item parameters for each group are displayed as circles in the three panels in the lower part of Figure 10.1, where the x-axis indicates the item number and the y-axis indicates the item difficulty.

The Rasch tree method was first suggested for dichotomous Rasch models, but has been extended to polytomous models, where it can detect differential step functioning (Komboz, Strobl, & Zeileis, 2017). It is available in the R package `psychotree` (Zeileis, Strobl, Wickelmaier, Komboz, & Kopf, 2021).

10.2.3 Explanatory IRT – The Rasch Model as a Mixed Model

Another approach for describing differences between test takers (including DIF) views the Rasch model as a mixed model, i.e., a model containing both "fixed" and "random" effects. You can find an introduction to mixed models in Gelman and Hill (2006) or Pinheiro and Bates (2000).

The difference between fixed and random effects can be understood by example. Suppose we are interested in whether different teaching methods influence final exam scores. Fixed effects would be used to account for the effects of different teaching methods on performance. Random effects would be used to account for covariance among observations. When modeling the effect of teaching method, it may be natural to assume at first sight that the measurements from individual students are independent of each other. However, if the students are nested within different classes, it should be taken into account that measurements of the students from the same class may be more similar than measurements of students from different classes, for example, because they are taught by the same teacher and experience similar conditions in the classroom. The differences between the measurements of the students from different classes can be described by means of random effects. The variance of the random effects will then represent the extent to which the measurements of the different classes vary. Modeling the differences between the classes as random effects also reflects the idea that the classes are typically not picked by design, like the teaching methods, but represent a random sample from a larger population of classes.

Sometimes, we have multiple levels of nesting: Students are nested in classes, which are nested in schools, which are nested in school districts, etc. Mixed models can be used to account for these relationships as well. In this case, each student is regarded as a random sample from their class, each class is regarded as a random sample from its school, each school is regarded as a random sample from its school district, etc. Based on this idea, mixed models are also called hierarchical or multilevel models. There are two major advantages of employing mixed models. First, so long as schools and classes are sampled randomly, we can generalize the influence of differences in these random factors beyond the sample. Second, we can incorporate covariates at different levels of the model.

The Rasch model can also be represented as a mixed model (De Boeck & Wilson, 2004; Rijmen, Tuerlinckx, De Boeck, & Kuppens, 2003). Test items are typically regarded as fixed, so their difficulties are estimated as fixed effects. Test takers are typically regarded as a random sample from a population, so their abilities are estimated as random effects. However, it is also possible to treat both test items and test takers as random (De Boeck, 2008). This method allows the Rasch model to be fit with software developed for mixed models. Additionally, it allows us to take hierarchical structure into account.

A framework that views the Rasch model as a mixed model is explanatory item response modeling or explanatory IRT (De Boeck & Wilson, 2004). In explanatory IRT, mixed models are used to understand how IRT model parameters, like ability and difficulty, differ over covariates, such as gender or age. In this framework, testing for DIF corresponds to checking whether an item parameter is stable over person covariates by incorporating the interaction of item and person variables into the model as a random effect (Rijmen et al., 2003). Mixed models can be estimated in R using the `lme4` package (for fitting IRT models with `lme4`, see De Boeck et al., 2011; for the information about the R package itself, see Bates, Mächler, Bolker, & Walker, 2015; Bates et al., 2022). They can also be estimated in Stan either from scratch or by using the R package `rstanarm` (Stan Development Team, 2020).

10.3 Multidimensional IRT Models

Another way of extending the Rasch model involves relaxing the unidimensionality assumption. Many tests consist of several parts. For example, the SAT is composed of both verbal and mathematical items. While some test takers may be equally good at both parts, there will be many people who are much better at one part than the other. It therefore makes sense to treat the two parts as separate dimensions and to report their scores separately. We could later combine the sections into a single aggregate score, but this would require some form of weighting of the two sub-scores.

Multidimensional IRT models are used to account for tests with multidimensional items. In multidimensional IRT models, each person's test-taking behavior is determined by multiple latent traits or factor scores. The term factor score derives from factor analysis. Multidimensional IRT and factor analysis are closely related (e.g., Reckase, 2009). As a result, many ideas and a considerable amount of terminology have been adopted into multidimensional IRT from factor analysis.

In multidimensional IRT models, the response probabilities for each item depend on one or more of the latent traits. In the simplest case, every item depends, or loads, on exactly one of the person's factor scores. This was the case in our SAT example, where every item was either a verbal item or a math item, but never both. The case where every item loads on exactly one factor is often called "between-item multidimensionality" or "simple structure". An item can also load on multiple factors. This is known as "within-item multidimensionality" (Hartig & Höhler, 2008).

In R, multidimensional IRT models can be estimated with both the `mirt` (see also Desjardins & Bulut, 2018) and `TAM` packages. An alternative is using Bayesian inference in Stan.

10.4 Exercises

1. What is the meaning of the weights w_{ia} in the linear-logistic test model?
2. What is the main conceptual difference between mixture Rasch models on the one hand and model-based recursive partitioning and explanatory Rasch models on the other hand?
3. Is it possible to create a joint score from two separate dimensions, such as verbal and mathematical reasoning?

11

Models for Polytomous Responses

CONTENTS

Up to now, we have focused on models for dichotomous items, where a test taker either solves or agrees with the item (coded as 1) or does not (coded as 0). Polytomous Rasch and IRT models extend Rasch and 2PL models to account for items with more than two response categories, which are very common in many areas of educational and psychological testing. Examples include ability tests giving partial credit for partially correct answers and personality or attitude scales asking for different levels of agreement.

For didactic reasons, we will present these models as mathematical generalizations of the Rasch model, but note again that historically some of them were developed independently and not necessarily in the order in which they are presented here. In this chapter, we will touch upon a variety of polytomous models to point out how they relate to each other, but you will notice that for three models we provide a few more details. These details were mainly chosen to emphasize the fact that polytomous models require additional modeling decisions compared to dichotomous ones. The three models that we treat in more depth are the Partial Credit model, for which we provide the most details, the Rating Scale model, and the Graded Response model.

Readers interested in applying polytomous models are pointed to the respective R packages and additional literature at the end of each section. A more thorough treatment of polytomous models is provided in, e.g., Thissen and Steinberg (1986), van der Linden and Hambleton (1997), De Ayala (2009),

Penfield (2014), or Desjardins and Bulut (2018). There are also extensions of polytomous models for multidimensional traits (Reckase, 2009).

As we have already seen in Section 4.5.1.4, models including more parameters require larger sample sizes for estimation. This will also be discussed for the polytomous models we have presented in detail in Section 11.6.

11.1 The Partial Credit Model

We will first turn to two polytomous models from the Rasch family, the Partial Credit model and the Rating Scale model. The Partial Credit model (Masters, 1982) has been presented in different ways by different books and articles, and, at first glance, these presentations can appear to have little in common. To help the reader better understand this model, we will explain a few of the different ways that the Partial Credit model has been presented and show how they are related to one another.

Suppose that item i has $C_i + 1$ response categories labelled $0, 1, \ldots, C_i$. For example, an item on a personality test might have response categories "disagree", "neutral", and "agree". This item would have three categories (0, 1, and 2, with the highest category $C_i = 2$ and the overall number of categories being $C_i + 1 = 3$) and labels "disagree" for category 0, "neutral" for category 1, and "agree" for category 2. Alternatively, an item on an ability test might be graded on a four point scale, with 0 points awarded for an entirely incorrect response and 3 points awarded for an entirely correct response. The remaining point values, 1 and 2, are assigned as partial credit for increasingly correct responses. In this case, the highest category $C_i = 3$ and the overall number of categories $C_i + 1 = 4$. In the following, we will keep referring to the latent trait as an ability, but the same principles apply to personality or attitude scales.

For describing the response probabilities for the different categories, polytomous models use more than one ICC. Remember how, for the dichotomous Rasch model with two response categories, we usually considered only the probability of correctly answering the item, $\Pr(U_{pi} = 1 \mid \theta_p, \beta_i)$. The typical ICC for the Rasch model is the graphical representation of this probability as a function of the test taker's ability. However, it is also possible to look at the converse probability, i.e. the probability of not answering the item correctly $\Pr(U_{pi} = 0 \mid \theta_p, \beta_i) = 1 - \Pr(U_{pi} = 1 \mid \theta_p, \beta_i)$, and display this probability in the form of an ICC, too. The ICCs for both responses, zero and one, are displayed for the Rasch model in the exercises for the `TAM` package on p. 282.

For models with more than two response categories, this is carried forward. The specification of many polytomous models is based on the principle that the probabilities of responding in any individual category must sum up to one. As a result, these models have one curve per category, which are accordingly

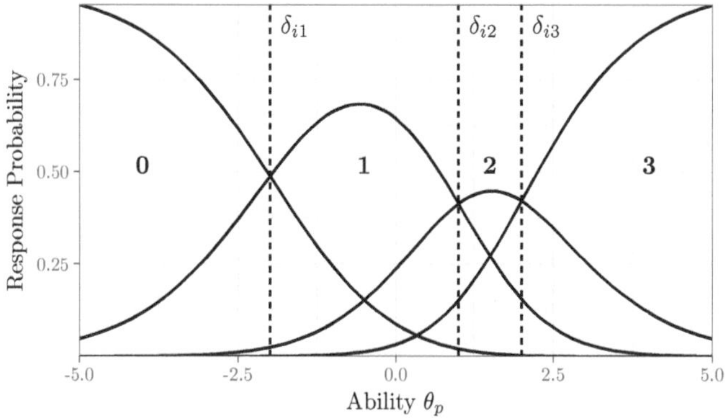

FIGURE 11.1: CCCs for an item with threshold parameters δ_{i1}, δ_{i2}, and δ_{i3}. The bolded numerals indicate the category with the highest response probability.

termed category characteristic curves (CCCs). For example, Figure 11.1 shows the CCCs for a Partial Credit model with response categories 0 through 3.

In this figure, the heights of the individual CCCs display the probabilities for answering in each category. The points where the probabilities of two adjacent categories are equal, δ_{i1}, δ_{i2}, and δ_{i3}, are termed thresholds and are indicated by dashed lines in Figure 11.1. The thresholds are model parameters that describe the location and spread of the items. Test takers with an ability up to the first threshold have the highest probability of answering with category 0, test takers with an ability between the first and second threshold have the highest probability of answering with category 1, etc. For any particular value of the latent trait, the response probabilities sum to one. For example, a person situated at the second threshold has a probability of about 40% each for answering with category 1 or category 2, a probability of around 15% for answering with category 3, and a probability around 5% for answering with category 0. These four probabilities add up to one.

Mathematically, the Partial Credit model builds the probabilities of the response categories from the ratios between the probabilities of adjacent response categories. For simplicity, let us consider the even simpler case where $C_i = 2$, so that we have two pairs of adjacent response categories: $\{0, 1\}$ and $\{1, 2\}$. The Partial Credit model defines the ratio between the probabilities of responding with category 1 and of responding with category 0 to be

$$\frac{\Pr(U_{pi} = 1 \mid \theta_p, \delta_{i1}, \delta_{i2})}{\Pr(U_{pi} = 0 \mid \theta_p, \delta_{i1}, \delta_{i2})} = \exp(\theta_p - \delta_{i1}), \tag{11.1}$$

and the ratio between the probabilities of responding with category 2 and of responding with category 1 to be

$$\frac{\Pr(U_{pi} = 2 \mid \theta_p, \delta_{i1}, \delta_{i2})}{\Pr(U_{pi} = 1 \mid \theta_p, \delta_{i1}, \delta_{i2})} = \exp(\theta_p - \delta_{i2}). \tag{11.2}$$

Taking the log of both sides of the first equation yields, for example,

$$\log \left\{ \frac{\Pr(U_{pi} = 1 \mid \theta_p, \delta_{i1}, \delta_{i2})}{\Pr(U_{pi} = 0 \mid \theta_p, \delta_{i1}, \delta_{i2})} \right\} = \theta_p - \delta_{i1}.$$

This expression is nearly identical to the logit form of the Rasch model in Equation (2.2). The only difference between the two equations is that the β_i in the Rasch model has been replaced with δ_{i1}, or respectively δ_{i2} for the second equation. Recall that in the Rasch model the item's difficulty β_i determined the ability at which the probability of responding correctly is the same as the probability of responding incorrectly. The role of δ_{i1} and δ_{i2} in the Partial Credit model is analogous to the role of β_i in the Rasch model. The first threshold parameter δ_{i1} determines the ability for which the probability of responding with 0 is the same as the probability of responding with 1, and the second threshold parameter δ_{i2} determines the ability for which the probability of responding with 1 is the same as the probability of responding with 2.

We can work out an expression for the probability that a person responds with category c in the case where $C_i = 2$ by solving a system of three equations. The first two equations are Equation (11.1) and Equation (11.2). The third states that the sum of the three probabilities is one. This equation comes from the fact that 0, 1, and 2 are the only possible responses to the item, so that their probabilities for any given value of θ_p must sum to one. Solving these equations for the three response probabilities[1] results in

$$\Pr(U_{pi} = 0 \mid \theta_p, \delta_{i1}, \delta_{i2}) = \frac{1}{1 + \exp(\theta_p - \delta_{i1}) + \exp\{(\theta_p - \delta_{i1}) + (\theta_p - \delta_{i2})\}} \tag{11.3}$$

for response category 0,

$$\Pr(U_{pi} = 1 \mid \theta_p, \delta_{i1}, \delta_{i2}) = \frac{\exp(\theta_p - \delta_{i1})}{1 + \exp(\theta_p - \delta_{i1}) + \exp\{(\theta_p - \delta_{i1}) + (\theta_p - \delta_{i2})\}} \tag{11.4}$$

for response category 1, and

$$\Pr(U_{pi} = 2 \mid \theta_p, \delta_{i1}, \delta_{i2}) = \frac{\exp\{(\theta_p - \delta_{i1}) + (\theta_p - \delta_{i2})\}}{1 + \exp(\theta_p - \delta_{i1}) + \exp\{(\theta_p - \delta_{i1}) + (\theta_p - \delta_{i2})\}} \tag{11.5}$$

for response category 2.

[1] Derivations for the equations presented in this section are collected at the end of this chapter because they are instructive but take up some space.

Notice how each progressive increase in response category incorporates an additional $\theta_p - \delta_{ic}$ term in the numerator. We start with $\Pr(U_{pi} = 0 \mid \theta_p, \delta_{i1}, \delta_{i2})$, where there is no $\theta_p - \delta_{ic}$ term visible in the numerator.[2] In $\Pr(U_{pi} = 1 \mid \theta_p, \delta_{i1}, \delta_{i2})$, we incorporate a single $\theta_p - \delta_{ic}$ term, namely $\theta_p - \delta_{i1}$. In $\Pr(U_{pi} = 2 \mid \theta_p, \delta_{i1}, \delta_{i2})$, we incorporate a second $\theta_p - \delta_{ic}$ term, namely $\theta_p - \delta_{i2}$, in addition to the first. Also note that the common denominator for each term is the sum of the three numerators, because the three probabilities add up to one.

Now that we understand the specific case where $C_i = 2$, we can derive the distribution of response probabilities for arbitrary C_i. As in the case where $C_i = 2$, we model the log probability ratios for adjacent response categories. This gives C_i log ratios,

$$\log\left\{\frac{\Pr(U_{pi} = c \mid \theta_p, \delta_{ic})}{\Pr(U_{pi} = c - 1 \mid \theta_p, \delta_{ic})}\right\} = \theta_p - \delta_{ic},$$

for $c = 1, \ldots, C_i$. Thus, we have C_i threshold parameters δ_{ic}, which we collect into a vector $\boldsymbol{\delta}_i = (\delta_{i1}, \ldots, \delta_{iC_i})$. We combine the log probability ratio equations with the fact that the probabilities for all response categories sum to 1, $\sum_{c=0}^{C_i} \Pr(U_{pi} = c \mid \theta_p, \delta_{ic}) = 1$, to determine that the probability that a test taker responds with category c is

$$\Pr(U_{pi} = 0 \mid \theta_p, \boldsymbol{\delta}_i) = \frac{1}{1 + \sum_{l=1}^{C_i} \exp\{\sum_{k=1}^{l}(\theta_p - \delta_{ik})\}} \tag{11.6}$$

when $c = 0$ and

$$\Pr(U_{pi} = c \mid \theta_p, \boldsymbol{\delta}_i) = \frac{\exp\{\sum_{k=1}^{c}(\theta_p - \delta_{ik})\}}{1 + \sum_{l=1}^{C_i} \exp\{\sum_{k=1}^{l}(\theta_p - \delta_{ik})\}} \tag{11.7}$$

when $c \geq 1$. You might note that in these equations the category index c is replaced by k or l in some places. While c denotes the category of interest, for which we are computing the probability in Equation (11.7), k and l are indices representing the categories currently used in each step of the respective sums.

The Partial Credit model reduces to the Rasch model when there are only two response categories, 0 and 1. We can see this from Equation (11.7) by setting $C_i = 1$ and $\delta_{i1} = \beta_i$.

11.1.1 CCCs and Threshold Parameters

As already outlined above, for the Partial Credit model we plot $C_i + 1$ individual CCCs, each of which showing the probability of responding with category c as a function of ability.

---

[2]Here we can think of the term in the numerator as a sum over zero terms. A sum over zero terms is 0 by definition. Thus we get $\exp(0) = 1$ (see Appendix A) for the numerator.

In Figure 11.1 we have shown the CCCs and corresponding threshold parameters δ_{ic} of the Partial Credit model for an item with four response categories. These parameters are also called Rasch-Andrich thresholds (Bond & Fox, 2007). Scanning from left to right, we have seen that a person whose ability is below δ_{i1} is most likely to respond with category 0, a person whose ability is between δ_{i1} and δ_{i2} is most likely to respond with category 1, etc.

We can also see the shapes of the different CCCs. The CCC for the lowest category 0 is always decreasing in ability. By contrast, the CCCs for categories 1 and 2 look like bells: they increase in ability until they hit their maximum, after which they decrease in increasing ability. This maximum falls between δ_{i1} and δ_{i2} for response category 1 and between δ_{i2} and δ_{i3} for response category 2. Finally, the CCC for the highest category, which is category 3 here, is always increasing in increasing ability. This pattern will hold for any item and ensures that the response probabilities sum up to one for each value of the latent trait.

Some software packages, for instance Winsteps (Linacre, 2020), condense the information contained in CCCs by providing additional threshold parameters, which differ from the Rasch-Andrich thresholds. Given a specific response category, these threshold parameters present the level of ability for which the probability of observing a response in this category or a lower category equals the probability of observing a response in the next category or a higher category (whereas the Rasch-Andrich thresholds refer to the level where the probabilities of two individual adjacent categories are equal). These alternative thresholds can also be useful for the interpretation of the ability scales and are sometimes called Thurstone thresholds in the literature. For a discussion of these thresholds in the context of various Rasch models, see, for instance, Bond and Fox (2007).

11.1.2 Alternative Parameterizations

Besides Thurstone thresholds, that have a different meaning and location, there are also two additional parameterizations of the Partial Credit model that are based on conversions of the Rasch-Andrich thresholds. Converting one type of parameter into another is called reparameterization. The values of the parameters do not have to be re-estimated from the data. Instead, there is a unique transformation which enables us to switch between the various parameterizations. This does, however, change the look of the CCC equations, and the model in its new formulation may be hard to recognize at first glance.

One common parameterization of the Partial Credit model comes from using the sums of the category thresholds as model parameters. In this, the parameters are $\lambda_{ic} = \sum_{k=1}^{c} \delta_{ik}$. Under this parameterization, for $c \geq 1$, we can use the fact that $\sum_{k=1}^{c} \theta_p = c \cdot \theta_p$ (which follows from a rule given in Appendix A), and obtain:

$$\exp\left\{\sum_{k=1}^{c}(\theta_p - \delta_{ik})\right\} = \exp\left\{c \cdot \theta_p - \sum_{k=1}^{c} \delta_{ik}\right\} = \exp\{c \cdot \theta_p - \lambda_{ic}\}.$$

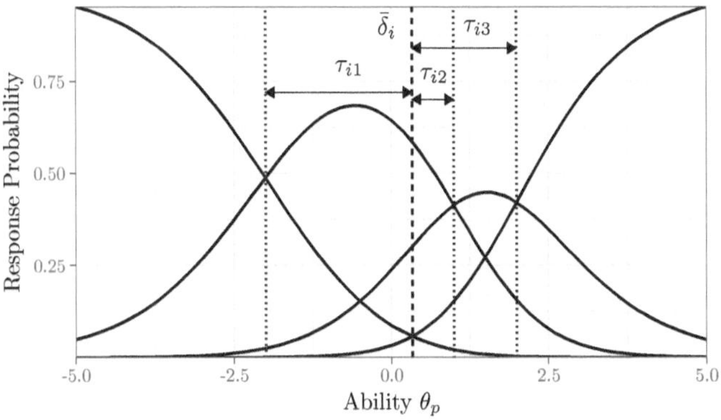

FIGURE 11.2: Illustration of the relationship between the threshold parameterization of the Partial Credit model and the parameterization using $\bar{\delta}_i$ and $\boldsymbol{\tau}_i$.

Defining $\boldsymbol{\lambda}_i$ to be the vector of λ_{ic} parameters, Equation (11.6) becomes

$$\Pr(U_{pi} = 0 \mid \theta_p, \boldsymbol{\lambda}_i) = \frac{1}{1 + \sum_{l=1}^{C_i} \exp(l \cdot \theta_p - \lambda_{il})}$$

and Equation (11.7) becomes

$$\Pr(U_{pi} = c \mid \theta_p, \boldsymbol{\lambda}_i) = \frac{\exp(c \cdot \theta_p - \lambda_{ic})}{1 + \sum_{l=1}^{C_i} \exp(l \cdot \theta_p - \lambda_{il})}.$$

This parameterization may be perceived as notationally simpler than the threshold parameterization by some. However, the λ_{ic} parameters are more difficult to interpret, making them less common in practical applications.

Another parameterization of the Partial Credit model comes from centering the original threshold parameters. To do this, we first compute the average of the threshold parameters, $\bar{\delta}_i = (1/C_i) \cdot \sum_{c=1}^{C_i} \delta_{ic}$. This average $\bar{\delta}_i$ describes the general location of the item. We then set $\tau_{ic} = \bar{\delta}_i - \delta_{ic}$. These differences τ_{ic} describe the distances of the individual thresholds from the general location of the item. Now define $\boldsymbol{\tau}_i$ to be the vector of τ_{ic} parameters. Then, Equation (11.6) becomes

$$\Pr(U_{pi} = 0 \mid \theta_p, \bar{\delta}_i, \boldsymbol{\tau}_i) = \frac{1}{1 + \sum_{l=1}^{C_i} \exp\{\sum_{k=1}^{l} [\theta_p - (\bar{\delta}_i - \tau_{ik})]\}} \qquad (11.8)$$

and Equation (11.7) becomes

$$\Pr(U_{pi} = c \mid \theta_p, \bar{\delta}_i, \boldsymbol{\tau}_i) = \frac{\exp\{\sum_{k=1}^{c}[\theta_p - (\bar{\delta}_i - \tau_{ik})]\}}{1 + \sum_{l=1}^{C_i} \exp\{\sum_{k=1}^{l}[\theta_p - (\bar{\delta}_i - \tau_{ik})]\}}. \tag{11.9}$$

Figure 11.2 replots Figure 11.1 to illustrate the "centered" parameterization of the Partial Credit model. In both figures, the CCCs are in the same places, but the threshold positions are indicated by different sets of parameters. We see that $\bar{\delta}_i$ gives the general location of the item, i.e. the center of the thresholds, and that each τ_{ic} gives the offset or distance of the threshold for category c from the general location. This way of parameterizing the Partial Credit model will prove useful in comparing the model to another polytomous response model, the Rating Scale model, in Section 11.2.

Unfortunately, it can also be somewhat misleading. Thinking of $\bar{\delta}_i$ as the location of an item could mislead one to believe that an item i is easier than an item j if $\bar{\delta}_i < \bar{\delta}_j$. However, this forgets the fact that the distances between the thresholds are also free to vary across the two items, as illustrated by Figure 11.3.

Figure 11.3 shows the threshold parameters δ_{ic} and their mean $\bar{\delta}_i$ for three different items, shown one item per panel. Comparing the average threshold $\bar{\delta}_i$ across the three panels, we see that $\bar{\delta}_i = 0$ for the upper panel and $\bar{\delta}_i = 1$ for the lower and middle panels. Comparing the individual threshold parameters, we see that every threshold parameter in the middle panel falls to the right of the corresponding threshold parameter in the upper panel, but that one threshold parameter in the lower panel does not. Thus, it makes sense to say that the middle item is more difficult than the upper item, because every point of partial credit is harder to obtain. By contrast, it does not make sense to say that the lower item is more difficult than the upper item. A lower ability is needed in the lower item to get one point as compared to the upper item, but a higher ability is needed to get two or three points.

We can see this even more clearly by considering expected scores. The expected score for a certain person or a given ability can be computed by first multiplying each possible score, $0, \ldots, C_i$, by the probability of receiving this score given the ability. The sum over these results is the expected score. For any person or given ability value, we can determine the probability of scoring in category c by finding the probability corresponding to their ability on the corresponding CCC.

Let's compare the expected scores for the middle and lower items for a person whose ability $\theta_p = -3$. Looking at Figure 11.3 we see that the probabilities (heights of CCCs) of scoring two or three points for a person with ability $\theta_p = -3$ (on the x-axis), $\Pr(U_{pi} = 2 \mid \theta_p = -3, \boldsymbol{\delta}_i) \approx 0$ and $\Pr(U_{pi} = 3 \mid \theta_p = -3, \boldsymbol{\delta}_i) \approx 0$, are very close to zero for both items. Moreover, we can read off the probabilities of scoring zero or one points for each item. These probabilities are $\Pr(U_{pi} = 0 \mid \theta_p = -3, \boldsymbol{\delta}_i) \approx 0.85$ and $\Pr(U_{pi} = 1 \mid \theta_p = -3, \boldsymbol{\delta}_i) \approx 0.15$ for the middle item and $\Pr(U_{pi} = 0 \mid \theta_p = -3, \boldsymbol{\delta}_i) \approx 0.56$

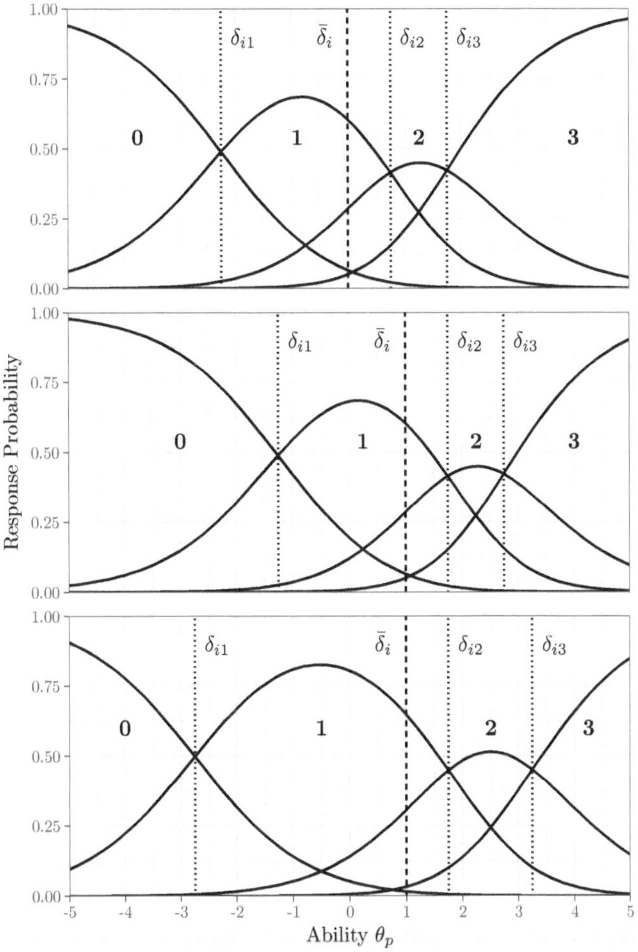

FIGURE 11.3: CCCs for different values of $\boldsymbol{\delta}_i$. The dashed lines show $\bar{\delta}_i$ and the dotted lines show the individual δ_{ic}.

and $\Pr(U_{pi} = 1 \mid \theta_p = -3, \boldsymbol{\delta}_i) \approx 0.44$ for the lower item. Thus, the expected score for the middle item is

$$\begin{aligned} E(U_{pi} \mid \theta_p = -3, \boldsymbol{\delta}_i) &= \sum_{c=0}^{C_i} c \cdot \Pr(U_{pi} = c \mid \theta_p = -3, \boldsymbol{\delta}_i) \\ &\approx 0 \times 0.85 + 1 \times 0.15 + 2 \times 0 + 3 \times 0 \\ &= 0.15 \end{aligned}$$

and the expected score for the lower item is 0.44, respectively. Thus, the expected score for a person with an ability of -3 for the lower item is higher than for the middle item despite the fact that the average threshold parameter $\bar{\delta}_i$ is the same for both items.

11.1.3 Disordered Thresholds

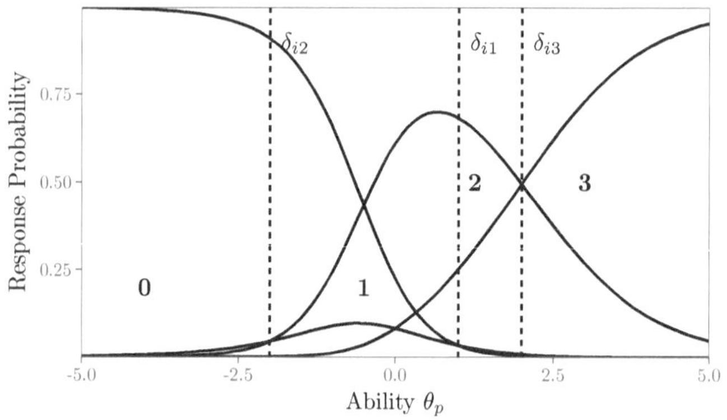

FIGURE 11.4: CCCs for an item with disordered category threshold parameters.

When working with the Partial Credit model, we expect that increasing ability will result in higher test scores. However, it turns out that this is not required by the Partial Credit model. Figure 11.4 shows the CCCs for an item whose threshold parameters are not ordered in the expected way. From the shape of the CCCs for this item, we would expect very few people to score 1 point. Since the probability for this category is always lower than the probabilities for the other categories, the order of the threshold parameters δ_{i1} and δ_{i2} is reversed. This makes their interpretation as thresholds on the latent scale implausible.

When we find such disordered thresholds for an item, there are different opinions on how they should be interpreted. Andrich (2010, as well as previous works) argues that disordered thresholds indicate that the response categories of the item in question were not used by the test takers in the ordinal way intended by the test designers. This point of view, shared by many authors (e.g., Rost, 1990), suggests that the Partial Credit model can be used to test whether response categories were used in the intended way. Adams, Wu, and Wilson (2012), however, argue against this view, referring to two definitions of order that the Partial Credit model does fulfill even when the threshold parameters are disordered.

What all agree on is that disordered threshold parameters indicate that the response categories in question rarely occur in the data. This is often the result of poor formulation of the response categories or the evaluation scheme, as illustrated by van Wyke and Andrich (2006) for the evaluation of geometric drawings. Third graders were given an illustration with several daisies on it. The task was to form a rectangle with the aid of a ruler by connecting some of the daisies. According to the evaluation scheme, 0 points were to be awarded if the response was wholly inadequate, 1 point if the correct daisies had been selected but the rectangle was poorly drawn (e.g., with sides that were not straight or parallel) and 2 points if the rectangle was entirely accurate. In practice, the middle category (1 point) was probably hardly ever used because, as van Wyke and Andrich argue, the children either understood the concept of a rectangle (in which case they did not make any mistakes when drawing it with the ruler and scored 2 points) or they did not understand it (which also meant that they were unable to pick the right daisies that would have been required for 1 point and consequently scored 0 points).

We can also encounter this problem in personality or attitude scales. To avoid having rarely or unused categories, it is particularly important to conduct a pretest with representative participants when incorporating items with polytomous responses. If the data has already been collected, the standard suggestion is to merge the category in question with one or more of the adjacent categories. This requires recoding the data and re-estimating the Partial Credit model. That said, Adams et al. (2012) and Wetzel and Carstensen (2014) advise against routinely merging response categories in the presence of disordered threshold parameters.

11.2 The Rating Scale Model

The Rating Scale model (Andrich, 1978) is a special case of the Partial Credit model. It is only applicable to – but not necessarily empirically fulfilled by – items sharing a common response format. For example, a questionnaire whose items all share the response categories “disagree”, “neutral”, and “agree” is a candidate for the Rating Scale model. By contrast, a questionnaire that contains a mix of items, some with these three response categories and others with five response categories, such as “strongly disagree”, “disagree”, “neutral”, “agree”, and “strongly agree”, is not.

The Rating Scale model fixes the number of response categories C_i and the threshold distances $\boldsymbol{\tau}_i$ to be the same value for every item. Let C be the common number of response categories and let every item share the same threshold distances $\boldsymbol{\tau} = \tau_1, \ldots, \tau_C$. Then, the centered parameterization of Equation (11.8) becomes

$$\Pr(U_{pi} = 0 \mid \theta_p, \bar{\delta}_i, \boldsymbol{\tau}) = \frac{1}{1 + \sum_{l=1}^{C} \exp\{\sum_{k=1}^{l} [\theta_p - (\bar{\delta}_i - \tau_k)]\}}$$

and Equation (11.9) becomes

$$\Pr(U_{pi} = c \mid \theta_p, \bar{\delta}_i, \boldsymbol{\tau}) = \frac{\exp\{\sum_{k=1}^{c}[\theta_p - (\bar{\delta}_i - \tau_k)]\}}{1 + \sum_{l=1}^{C} \exp\{\sum_{k=1}^{l}[\theta_p - (\bar{\delta}_i - \tau_k)]\}}. \quad (11.10)$$

The main difference to Equation (11.8) and Equation (11.9) is that for the Rating Scale model the threshold distances τ_c no longer have an index for the item i. This highlights that they are the same for all items. The only parameter that is specific to every item is the location parameter $\bar{\delta}_i$. This means one item can be further left or right of another, depending on their $\bar{\delta}_i$, but the intervals between thresholds are the same for all items, as for the first two items in Figure 11.3. This allows us to say that items are ordered with respect to their difficulty (Sijtsma & Hemker, 1998).

Just as with the Partial Credit model, the threshold parameters in the Rating Scale model can in principle be disordered. However, since the threshold distances τ_c are averaged across all of the test items, we rarely observe this misordering in the overall τ_c parameters in the Rating Scale model. This also means that unintended uses of the response scale or inappropriate evaluation schemes in individual items may go unnoticed more easily than with the Partial Credit model.

The decision between the Partial Credit and Rating Scale models is usually made on the basis of characteristics of the test as well as statistical criteria. When the number of response categories differs between items, the Rating Scale model is not applicable. When the number of response categories is the same for all items, which model is appropriate depends on whether empirically the usage of the response categories is comparable across items or not. We can employ information criteria to decide whether the Rating Scale model is sufficient or whether the more general Partial Credit model is required. As was outlined in Section 4.5.3, information criteria balance model fit against complexity. In the case of Partial Credit vs. Rating Scale models, the Partial Credit model with its item-specific threshold distances has a much higher number of parameters, making it much more complex but also much more flexible than the Rating Scale model.

Both Partial Credit and Rating Scale models belong to the Rasch family of models, providing sufficient statistics for the person parameters. This allows their item parameters to be estimated by means of the conditional maximum likelihood approach. Accordingly, in R these models are available in the `eRm` package, which employs conditional maximum likelihood. They can also be estimated by means of marginal maximum likelihood in `mirt` and `TAM`, or with Bayesian approaches.

11.3 The Generalized Partial Credit and the Nominal Response Model

For the dichotomous models presented in the first part of the book, we have shown that, mathematically, the 2PL model is an extension of the Rasch model. The 2PL allows the slope or discrimination to differ between items. The same kind of extension is also possible for polytomous models. The Generalized Partial Credit model (Muraki, 1992) is an extension of the Partial Credit model, in which the discrimination can differ between items. Just like for dichotomous items, an item with a higher discrimination distinguishes more sharply between the response categories for similar values of the latent trait. Looking back at Figure 11.3, a higher discrimination would result in steeper CCCs for the first and last categories and more narrow CCCs for the intermediate categories of an item.

The Generalized Partial Credit model allows the discriminations to vary between items, but holds them constant between the response categories of the same item. A further generalization from the mathematical point of view (while actually being a predecessor historically) is provided by the Nominal Response model (Bock, 1972), which allows the discrimination to vary between categories, too. In R, Generalized Partial Credit and Nominal Response models can be estimated by means of marginal maximum likelihood in `mirt` and `TAM`, or with Bayesian approaches.

11.4 The Graded Response Model

All models mentioned so far have in common that they describe the probabilities of individual response categories in a certain way. They can be derived from the probabilities of adjacent categories, as has been shown in Section 11.1. The resulting formulas for the probabilities of individual response categories divide a term that is specific to the current response category by the total sum over all categories. This is why these types of models are also termed divide-by-total models (Thissen & Steinberg, 1986) or adjacent category models (Penfield, 2014; see also Agresti, 2002).

A mathematically different approach is that of the Graded Response model, which was developed very early in the history of IRT by Samejima (1969). This model describes the probability of an individual response category c as the difference between the probability for categories c or above and the probability for categories $c + 1$ or above. Based on this characteristic, it is termed a difference model by Thissen and Steinberg (1986) or a cumulative model by Penfield (2014). It is also a non-Rasch model, because it allows for different discrimination parameters between items.

We consider again a test item i with C_i+1 response categories. The number of categories can vary between items, like in the Partial Credit model. As usual, a response in a higher category is interpreted as indicating a higher value on the latent trait. For each response category c of item i, with $c = 0, \ldots, C_i$, the Graded Response model has an item and category specific boundary location parameter γ_{ic}, as well as an item specific discrimination parameter α_i. Since the category specific parameters of this model require a different interpretation than those of the Rasch-type models considered above, we use a different letter to denote them. Given these parameters and the usual person ability parameter θ_p, the Graded Response model now defines the probability of observing category c or higher as:

$$\Pr(U_{pi} \geq c \mid \theta_p, \alpha_i, \gamma_i) = \frac{\exp\{\alpha_i(\theta_p - \gamma_{ic})\}}{1 + \exp\{\alpha_i(\theta_p - \gamma_{ic})\}}.$$

Consequently, the probability for category c by itself is the difference of the cumulative probabilities for category c and category $c+1$, for with $c \leq C$:

$$\Pr(U_{pi} = c \mid \theta_p, \alpha_i, \gamma_i) = \Pr(U_{pi} \geq c \mid \theta_p, \alpha_i, \gamma_i) - \Pr(U_{pi} \geq c+1 \mid \theta_p, \alpha_i, \gamma_i)$$

For category $C+1$ we obtain:

$$\Pr(U_{pi} = C+1 \mid \theta_p, \alpha_i, \gamma_i) = \Pr(U_{pi} \geq C+1 \mid \theta_p, \alpha_i, \gamma_i)$$

A more thorough discussion of the Graded Response model is provided, e.g., by Embretson and Reise (2000) and De Ayala (2009). In R, the Graded Response model can be estimated by means of the `mirt` package.

11.5 The Sequential Model

A yet slightly different philosophy is that of the Sequential model (Tutz, 1990, 1997), which Penfield (2014) also refers to as a continuation ratio model. This model was especially intended to describe a scenario where reaching category c requires first reaching category $c-1$. This is often the case for mathematical items. For example, calculating $\sqrt{30-5}$ takes two steps. First, one must calculate $30 - 5 = 25$. Only then can the square root be taken, resulting in $\sqrt{25} = 5$. Tutz (1997) emphasizes that the steps in the Sequential model correspond to consecutive steps in problem solving. He focuses on a model with equal discrimination, but the sequential approach can be straightforwardly extended to include unequal discrimination. In R, the `mirt` package offers both variants. It can also be shown that the sequential approach of Tutz and the cumulative approach of Samejima's Graded Response model are equivalent in the special case when the extreme-value distribution function is used instead of the usual logistic form (Bechger & Akkermans, 2001; Tutz, 1997).

11.6 Sample Size Requirements

Like for the dichotomous models, De Ayala (2009) presents a review of the research literature on sample size requirements available at that time. These guidelines can serve as rough indicators of which sample sizes could be practically useful. We summarize this discussion here for those models we have presented in detail above, but point to the original work for a more extensive discussion of this issue.

What is important to note is that – besides the factors that were already discussed above for the dichotomous models, such as the number of items and the alignment and shape of the item and person parameter distributions – for polytomous models, the estimation accuracy also depends on the number of response categories, which greatly increases the overall number of item parameters, as well as their locations.

De Ayala (2009) reports several rules of thumb from the literature, including desirable ratios of persons to item parameters from 2:1 to 5:1. In summary, De Ayala (2009) suggests minimum sample sizes of about 250 for Partial Credit and Rating Scale models and 500 for Graded Response models (e.g., for 25 items with 5 categories).

Again, indications that the employed sample size was not sufficient for reliably estimating the model parameters are convergence problems and large standard errors.

11.7 Exercises

1. What property do Partial Credit and Rating Scale models share with Rasch models that allows their item parameters to be estimated by means of conditional maximum likelihood?
2. A partial credit item has four response categories (scored with 0, 1, 2, and 3). If you plotted the category characteristic curves for this item, how many of them would be bell-shaped? Why?
3. Suppose two professors have just given the same midterm exam for their introductory psychology courses. The test consists of 4 short answer items, which are graded on a scale of 0 to 5. Suppose that the first professor grades all of the items herself, but that the second professor has each of her four teaching assistants grade one item. Now suppose that only one of the two professors was able to use the Rating Scale model to grade her tests. Which professor do you think it was? Why?

4. Which estimation approaches are available for Graded Response and Generalized Partial Credit models?

11.8 Derivations for the Partial Credit Model

In the following we have collected the derivations for the central equations in Section 11.1 for the interested reader. First, the Partial Credit model equations will be derived for the special case with categories 0, 1, and 2, before addressing the general case with categories $0, 1, \ldots C_i$.

For the derivation of Equation (11.3) through Equation (11.5), let x_0, x_1, and x_2 be the probabilities that a person p responds with categories 0, 1, and 2, respectively. Equation (11.1) indicates that $x_1/x_0 = \exp(\theta_p - \delta_{i1})$. Multiplying both sides by x_0, we get $x_1 = \exp(\theta_p - \delta_{i1}) \cdot x_0$. Similarly, Equation (11.2) implies that $x_2 = \exp(\theta_p - \delta_{i2}) \cdot x_1$. We can substitute $\exp(\theta_p - \delta_{i1}) \cdot x_0$ for x_1, resulting in the equation $x_2 = \exp\{(\theta_p - \delta_{i1}) + (\theta_p - \delta_{i2})\} \cdot x_0$.

We can use the fact that $x_0 + x_1 + x_2 = 1$ to solve for x_0 by substituting $\exp(\theta_p - \delta_{i1}) \cdot x_0$ for x_1 and $\exp\{(\theta_p - \delta_{i1}) + (\theta_p - \delta_{i2})\} \cdot x_0$ for x_2. After collecting all of the terms, we get that

$$\left[1 + \exp(\theta_p - \delta_{i1}) + \exp\{(\theta_p - \delta_{i1}) + (\theta_p - \delta_{i2})\}\right] \cdot x_0 = 1.$$

Dividing by $1 + \exp(\theta_p - \delta_{i1}) + \exp\left\{(\theta_p - \delta_{i1}) + (\theta_p - \delta_{i2})\right\}$, we get that

$$x_0 = \frac{1}{1 + \exp(\theta_p - \delta_{i1}) + \exp\left\{(\theta_p - \delta_{i1}) + (\theta_p - \delta_{i2})\right\}}.$$

Plugging this value into the equations for x_1 and x_2, we get that

$$x_1 = \frac{\exp(\theta_p - \delta_{i1})}{1 + \exp(\theta_p - \delta_{i1}) + \exp\left\{(\theta_p - \delta_{i1}) + (\theta_p - \delta_{i2})\right\}}$$

and

$$x_2 = \frac{\exp\left\{(\theta_p - \delta_{i1}) + (\theta_p - \delta_{i2})\right\}}{1 + \exp(\theta_p - \delta_{i1}) + \exp\left\{(\theta_p - \delta_{i1}) + (\theta_p - \delta_{i2})\right\}}.$$

Now we have derived the Partial Credit model equations for the special case with categories 0, 1, and 2, Equation (11.3) through Equation (11.5). Next, we will move on to the general case with categories $0, 1, \ldots C_i$. The derivation of the corresponding equations in Section 11.1, Equation (11.6) and Equation (11.7), follows the same principle we have shown for the special case above, but uses a more general notation.

Let $x_0, x_1, \ldots x_{C_i}$ be the probabilities of responding with category $0, 1, \ldots C_i$. We start by expressing $x_1, \ldots, x_{C_i}$ in terms of x_0. Using the same reasoning as above, we get an expression for the x_1 case, since $x_1 = \exp(\theta_p - \delta_{i1}) \cdot x_0$. For x_c, we substitute the expression for x_{c-1} into the expression relating x_c with x_{c-1}. This gives

$$x_c = \exp(\theta_p - \delta_{ic}) \cdot x_{c-1} = \exp(\theta_p - \delta_{ic}) \cdot [\exp(\theta_p - \delta_{i,c-1}) \cdot x_{c-2}].$$

We can continue substituting the expressions relating x_{c-k} and x_{c-k-1} until we finally obtain

$$x_c = \exp(\theta_p - \delta_{ic}) \cdots \exp(\theta_p - \delta_{i1}) \cdot x_0 = \exp\left\{\sum_{k=1}^{c}(\theta_p - \delta_{ik})\right\} \cdot x_0.$$

Now we combine this expression with the fact that $\sum_{l=0}^{C_i} x_l = 1$ to obtain

$$x_0 + \sum_{l=1}^{C_i} \exp\left\{\sum_{k=1}^{l}(\theta_p - \delta_{ik})\right\} \cdot x_0 = 1.$$

Since both terms involve x_0, just like above, we can factor this equation into

$$x_0 \cdot \left[1 + \sum_{l=1}^{C_i} \exp\left\{\sum_{k=1}^{l}(\theta_p - \delta_{ik})\right\}\right] = 1.$$

Dividing by $1 + \sum_{l=1}^{C_i} \exp\{\sum_{k=1}^{l}(\theta_p - \delta_{ik})\}$ yields

$$x_0 = \frac{1}{1 + \sum_{l=1}^{C_i} \exp\{\sum_{k=1}^{l}(\theta_p - \delta_{ik})\}}.$$

Substituting x_0 into the expressions for the remaining x_c, we get

$$x_c = \frac{\exp\{\sum_{k=1}^{c}(\theta_p - \delta_{ik})\}}{1 + \sum_{l=1}^{C_i} \exp\{\sum_{k=1}^{l}(\theta_p - \delta_{ik})\}}.$$

12

Outlook on Special Applications

CONTENTS

In this last chapter of the book, we would like to point the reader to a few application areas and study designs where the Rasch model and related models can be particularly helpful. Again, we will not go into much detail here, but point the interested reader to helpful references, both with respect to the underlying theory and to the practical application in R.

12.1 Computerized Adaptive Testing

Computerized adaptive testing (CAT) is an important application of IRT models. Up to now, we have considered the situation where the same test is administered to every test taker. Tests of this type are sometimes named linear tests to discern them from adaptive tests, which we will discuss here. We have demonstrated in Section 3.6 that tests constructed in this way will not provide the same information for every test taker. As a consequence, test takers who are near the center of the scale will typically have their abilities estimated more accurately than more extreme test takers near the ends.

Computerized adaptive testing solves this problem by administering different tests to different test takers. In computerized adaptive testing, each test is constructed at the time of administration from a large bank of items. The presented items are chosen to give as much information as possible about the ability of the test taker. More information about computerized adaptive testing can be found in van der Linden and Glas (2010) and Wainer et al. (2000).

To better understand how CAT works, suppose that we have an item pool of I items, from which we administered $\tilde{I}$ items with difficulties $\beta_1, \ldots, \beta_{\tilde{I}}$ to test taker p resulting in responses $U_{p1}, \ldots, U_{p\tilde{I}}$. We can use p's responses and the difficulties of the test items to get a provisional estimate $\hat{\theta}_{p\tilde{I}}$ of p's ability. We can then select the item maximizing the item information $\mathcal{I}(\hat{\theta}_{p\tilde{I}})$ for our provisional estimate as item $\tilde{I}+1$.

A requirement for computerized adaptive testing is that the item parameters need to be known in advance, since the item parameters are used both for generating provisional ability estimates and for selecting items from the item bank. IRT methods allow us to estimate these parameters using one group of test takers and subsequently use our estimates to score another set of test takers. This way test developers can construct large item banks to be used in computerized adaptive testing. In fact, the possibility to construct item banks is a key benefit of the Rasch and IRT frameworks over other test theory approaches such as classical test theory. A closely related method of test administration is multistage testing (Yan, von Davier, & Lewis, 2016). Here, groups of items are administered adaptively, instead of individual items.

Conceptually, the estimation of item and person parameters in the context of adaptive tests is closely related to the regular estimation of item and person parameters with missing data. Therefore, the R packages presented in this book can also be used in this context. In addition, there are packages for simulating and administering computerized adaptive and multistage testing: `catR` (Magis & Barrada, 2017; Magis & Raîche, 2012b), `mstR` (Magis, Yan, & von Davier, 2018), and `mirtCAT` (Chalmers, 2016).

12.2 Test Linking and Equating

Test linking and equating are another important application of IRT models. The main goal of these techniques can be summarized as follows: In many psychological and educational assessments, several forms of an assessment, consisting of different items, are used alongside each other. Naturally, these various test forms might differ in their relative difficulty. According to Kolen and Brennan (2014, p. 2), "Equating is a statistical process that is used to adjust scores on test forms so that scores on the forms can be used interchangeably."

A related, but different term is linking. According to Kolen and Brennan (2014), linking is applied to tests that aim to differ with regard to their difficulty, whereas equating is used with test forms that are designed to be as comparable as possible with regard to their content and statistical characteristics. We would like to add that these terms are not used consistently in the literature (von Davier, 2011b).

Although test linking and equating do not strictly require the application of Rasch or IRT models, using these models can be very helpful. Typically, their application leads to the aim of placing the item parameters from both test forms on a common scale (von Davier, 2011b). This can be done, for instance, by using a set of anchor items which is shared by both test forms. The resulting item parameters can then be readily estimated, e.g., using the R packages presented in the second part of the book. A check of the assumption that the item parameters of the anchor items are invariant for both populations leads back to the problem of the detection of DIF, which was addressed in the first part of the book. For an introduction to test linking and equating, see De Ayala (2009), Dorans, Pommerich, and Holland (2007), Kolen and Brennan (2014), or von Davier (2011a).

Besides the R packages already presented in this book, there are several R packages that provide methods for linking and equating in an IRT context. Among those are `equateIRT` (Battauz, 2015, 2021a), `equateMultiple` (Battauz, 2021b), and `kequate` (Andersson, Bränberg, & Wiberg, 2013). The R package `equate` (Albano, 2016) provides functions for equating without the use of IRT models.

12.3 Longitudinal IRT Models

For longitudinal studies and repeated measures, it is possible to use Rasch and IRT models to describe changes over time. One often used principle here is that changes in the person parameters over time will manifest as changes in the item parameters for the different measurement occasions when the person parameters are fixed over time. For example, if the persons increase in ability this will show as the items decreasing in difficulty.

Fischer (1995) formalizes this idea by considering an item administered at two time points as two virtual items. The parameters of these virtual items can then be conceived as a compound of item parameters and so-called treatment effects. Typically, it is assumed that for the first time point the item parameter reflects a baseline difficulty, while for following time points, it reflects the baseline difficulty plus any change in the ability, i.e., the treatment effect. In this conceptual framework, treatment effects can be used to model the effects of actual treatments, such as a medical treatment or an intervention administered between two time points, as well as changes over time in general. The concept of virtual items is further explained in Mair and Hatzinger (2007).

It is possible to model global and item-specific time effects by means of the linear-logistic test model introduced in Section 10.1 and its extensions in `eRm` by means of specifying appropriate design matrices. The theory behind this approach is described in detail by Fischer (1995) and its application in `eRm` is illustrated in detail by Mair (2018).

An alternative, which is also described in detail by Mair (2018), is to use the two-tier model of Cai (2010), which can be fit in `mirt` as a bifactor model. In his example, Mair (2018) uses two primary factors for the two time points, as well as specific factors to capture pairwise residual correlations. Yet other alternatives pointed out by Mair (2018) are latent growth IRT models, which can also be fit in `mirt`, as well as dynamic models estimated in a Bayesian framework.

It is also possible to combine longitudinal models with a mixture distribution approach (Meiser, 2007). This allows one to capture differences in the change over time between latent groups of test takers.

12.4 Exercises

1. Name two advantages that computerized adaptive testing has for test takers.
2. What is the role of anchors in test equating?
3. Explain the concept of virtual items in longitudinal IRT.

Appendices

A

Useful Mathematical Formulas

CONTENTS

A.1 Sums and Products

- $\sum_{p=1}^{P} x_p = x_1 + x_2 + \ldots + x_P$. For example, $\sum_{p=1}^{3} x_p = x_1 + x_2 + x_3$.
- $\sum_{p=1}^{P} a = P \cdot a$. For example, $\sum_{p=1}^{3} 2 = 2 + 2 + 2 = 3 \times 2 = 6$.
- $\sum_{p=1}^{P} a \cdot x_p = a \cdot \sum_{p=1}^{P} x_p$. For example,

$$\begin{aligned} \sum_{p=1}^{3}(2 \cdot x_p) &= 2 \cdot x_1 + 2 \cdot x_2 + 2 \cdot x_3 \\ &= 2 \cdot (x_1 + x_2 + x_3). \end{aligned}$$

- $\sum_{p=1}^{P}(a + x_p) = P \cdot a + \sum_{p=1}^{P} x_p$. For example,

$$\sum_{p=1}^{3}(2 + x_p) = (2 + x_1) + (2 + x_2) + (2 + x_3)$$

$$3 \times 2 + (x_1 + x_2 + x_3).$$

- $\prod_{p=1}^{P} x_p = x_1 \cdot x_2 \cdots x_P$. For example, $\prod_{p=1}^{3} x_p = x_1 \cdot x_2 \cdot x_3$.
- $\prod_{p=1}^{P} a = a^P$. For example, $\prod_{p=1}^{3} 2 = 2 \times 2 \times 2 = 2^3 = 8$.

A.2 Exponentials

- $e^x = \exp(x)$.
- $e^1 = \exp(1) = e$.
- $e^0 = \exp(0) = 1$.
- $e^{-x} = \exp(-x) = 1/e^x = 1/\exp(x)$.
- $e^{1/x} = \exp(1/x) = \sqrt[x]{e}$. For example, $e^{1/2} = \exp(1/2) = \sqrt{e}$.
- $e^x \cdot e^y = \exp(x) \cdot \exp(y) = e^{x+y} = \exp(x+y)$.
- $\prod_{p=1}^{P} e^{x_p} = \prod_{p=1}^{P} \exp(x_p) = e^{\sum_{p=1}^{P} x_p} = \exp(\sum_{p=1}^{P} x_p)$.
- $e^x/e^y = \exp(x)/\exp(y) = e^{x-y} = \exp(x-y)$.
- $(e^x)^y = [\exp(x)]^y = e^{xy} = \exp(xy)$.

A.3 Logarithms

We use $\log(x)$ to refer to the natural logarithm of x, which is often also denoted by $\ln(x)$. However, the following rules extend to logarithms with any basis, not only to the natural logarithm with basis e.

- $\log(e) = 1$.
- $\log(1) = 0$.
- $\log(e^x) = \log(\exp(x)) = x$.
- If $x > 0$ and $y > 0$: $\log(x \cdot y) = \log(x) + \log(y)$.
- If $x > 0$ and $y > 0$: $\log(x/y) = \log(x) - \log(y)$
- If $x > 0$: $\log(x^y) = y \cdot \log(x)$

A.4 Differentiation Rules

We use standard notation for derivatives. The derivative of the function $f(x)$ is denoted by $f'(x)$. The derivative with respect to x of the expression within the curly braces is denoted by $\frac{d}{dx}\{\cdot\}$. For example, $\frac{d}{dx}\{3x^2+2\}$ is the derivative with respect to x of $3x^2+2$.

A.4.1 Rules

- *Constants.* For any constant b,

$$\frac{d}{dx}\{b\} = 0.$$

- *Polynomials.*

$$\frac{d}{dx}\{x^a\} = a \cdot x^{a-1}.$$

- *Exponentials.*

$$\frac{d}{dx}\{e^x\} = \frac{d}{dx}\{\exp(x)\} = \exp(x) = e^x.$$

- *Logarithms.*

$$\frac{d}{dx}\{\log(x)\} = \frac{1}{x}.$$

- *Constant multipliers.* For any constant c,

$$\frac{d}{dx}\{c \cdot f(x)\} = c \cdot f'(x).$$

- *Sums.* For differentiable functions $f_1(x), f_2(x), \ldots, f_P(x)$,

$$\frac{d}{dx}\left\{\sum_{p=1}^{P} f_p(x)\right\} = \sum_{p=1}^{P} f_p'(x).$$

- *Products ("The Product Rule").* For differentiable functions $f(x)$ and $g(x)$,

$$\frac{d}{dx}\{f(x) \cdot g(x)\} = f'(x) \cdot g(x) + f(x) \cdot g'(x).$$

- *Quotients.* For differentiable functions $f(x)$ and $g(x)$,

$$\frac{d}{dx}\left\{\frac{f(x)}{g(x)}\right\} = \frac{f'(x) \cdot g(x) - f(x) \cdot g'(x)}{[g(x)]^2}.$$

- *Function composition ("The Chain Rule").* For differentiable functions $f(x)$ and $g(x)$,

$$\frac{d}{dx}\{f(g(x))\} = f'(g(x)) \cdot g'(x).$$

A.4.2 Examples

- *Polynomial functions.* We take the derivative of $f(x) = 3x^2 + 2x + 5$.

$$\begin{aligned}
f'(x) &= \frac{d}{dx}\left\{3x^2 + 2x + 5\right\} \\
&= \frac{d}{dx}\left\{3x^2\right\} + \frac{d}{dx}\left\{2x\right\} + \frac{d}{dx}\left\{5\right\} && \text{(sum rule)} \\
&= \frac{d}{dx}\left\{3x^2\right\} + \frac{d}{dx}\left\{2x\right\} + 0 && \text{(constant rule)} \\
&= 3 \cdot \frac{d}{dx}\left\{x^2\right\} + 2 \cdot \frac{d}{dx}\left\{x\right\} && \text{(constant multiplier)} \\
&= 3 \cdot (2x) + 2 \cdot 1 && \text{(polynomial rules)} \\
&= 6x + 2. && \text{(simplification)}
\end{aligned}$$

- *Chain rule.* We take the derivative of $f(x) = (3x^2)^4$.

$$f(x) = \underbrace{(\underbrace{3x^2}_{h(x)})^4}_{g(h(x))}, \text{ so } f'(x) = \frac{d}{dx}\{g(h(x))\} = \underbrace{4 \cdot (3x^2)^3}_{g'(h(x))} \cdot \underbrace{(6x)}_{h'(x)}.$$

- *Chain rule with the exponential function.* We take the derivative of $f(x) = \exp(ax + b)$.

$$\begin{aligned}
f'(x) &= \frac{d}{dx}\{\exp(ax + b)\} \\
&= \exp(ax + b) \cdot \frac{d}{dx}\{ax + b\} && \text{(chain rule)} \\
&= \exp(ax + b) \cdot a. && \text{(sum and constant rules)}
\end{aligned}$$

B

Statistical Background

CONTENTS

B.1 Statistical Estimation

Suppose we are interested in whether a coin is fair, for example because we want to play a game where we win a price when the coin comes up heads. We could test whether the coin is fair or not by flipping the coin many times and counting the number of heads. A fair coin has an equal probability of coming up heads or tails (0.5), so we expect a fair coin to come up heads on around half of the flips. More concretely, define k to be the total number of flips and X to be the number of heads. Since we are always flipping the same coin, it seems reasonable to say that every flip has the same probability π of landing on heads and the same probability $1 - \pi$ of landing on tails. A fair coin will have $\pi = 1 - \pi = 0.5$, so we expect $X \approx k/2$.

B.1.1 The Binomial Distribution

Let's think about the probability of observing three heads followed by two tails on five flips. The probability that the first flip comes up heads is π. Because the result of the second flip is independent of the result of the first, we can obtain the probability that both flips are heads by multiplying the probability that the first flip is heads by the probability that the second flip is heads. Thus, the probability that the first two flips come up heads is $\pi \cdot \pi = \pi^2$. We can continue this argument to obtain the probability that the first three flips come up heads, $\pi \cdot \pi \cdot \pi = \pi^3$.

The result of the fourth and fifth flips is tails, which comes up with probability $1 - \pi$. Since the fourth flip is independent of the first three flips, the probability that we observe three heads followed by one tails is $\pi^3 \cdot (1 - \pi)$. Similarly, the fifth flip is independent of the first four, so the probability of three heads followed by two tails is $\pi^3 \cdot (1 - \pi)^2$.

We can easily verify that any combination of three heads and two tails will have the same probability of occurring. Thus, the probability of observing heads on the first flip, followed tails on the second, heads on the third and fourth, and tails on the fifth is $\pi^3 \cdot (1 - \pi)^2$, as is the probability of flipping tails on the first two flips and heads on the last three, etc. This allows us to compute the probability of observing three heads and two tails, in any order, by multiplying $\pi^3 \cdot (1 - \pi)^2$ by the number of all possible ways of flipping three heads and two tails. Since there are ten ways of doing this, the desired probability is $10 \cdot \pi^3 \cdot (1 - \pi)^2$.

How did we determine that there are ten ways of flipping three heads and two tails? One way is to simply enumerate the different ways of flipping three heads and two tails. However, this method of counting is impractical for larger numbers of flips. Thankfully, it turns out that the number of different ways of flipping three heads on five flips is given by the binomial coefficient

$$\binom{5}{3} = \frac{5!}{3! \cdot 2!}.$$

The "!" in this expression is shorthand for factorial. Thus,

$$\binom{5}{3} = \frac{5!}{3! \cdot 2!} = \frac{5 \cdot 4 \cdot 3 \cdot 2 \cdot 1}{(3 \cdot 2 \cdot 1) \cdot (2 \cdot 1)} = 10$$

after simplifying.

Combining these results, we see that the probability that three flips are heads is

$$\Pr(X = 3 \mid \pi, k = 5) = \binom{5}{3} \cdot \pi^3 \cdot (1 - \pi)^2.$$

We can easily extend this to any number of heads x between 0 and 5. From the previous paragraph, we know that there are $\binom{5}{x}$ different ways of flipping x heads and $5 - x$ tails on five flips. The probability of the x heads is π^x and the probability of the $5 - x$ tails is $(1 - \pi)^{5-x}$. This results in the more general expression

$$\Pr(X = x \mid \pi, k = 5) = \binom{5}{x} \cdot \pi^x \cdot (1 - \pi)^{5-x}.$$

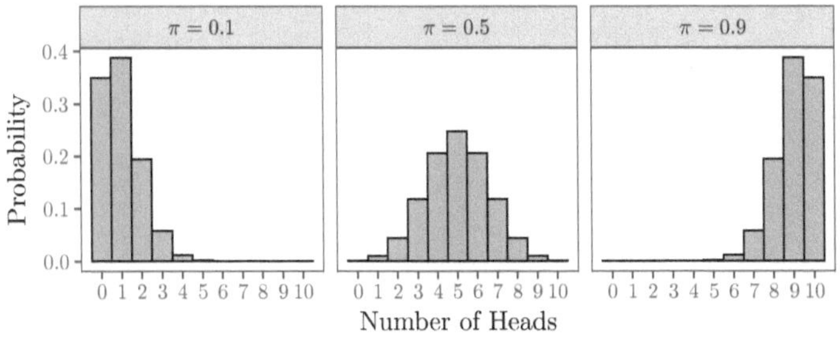

FIGURE B.1: $\Pr(X = x \mid \pi, k = 10)$ as a function of x for different values of π.

We can further generalize this by allowing the number of flips k to be different than 5. To do this, we simply repeat the previous arguments, replacing 5 with k. This results in the formula for the binomial distribution

$$\Pr(X = x \mid \pi, k) = \binom{k}{x} \cdot \pi^x \cdot (1 - \pi)^{k-x}.$$

We can best understand the binomial distribution by plotting its form for different values of π and k. Figure B.1 shows $\Pr(X = x \mid \pi, k = 10)$ as a function of x for $\pi = 0.1, 0.5$ or 0.9 for $k = 10$ flips. This plot demonstrates that the form of the distribution of X strongly depends on the value of π. When $\pi = 0.1$, the most likely number of heads on $k = 10$ flips is one, followed by zero and two. When $\pi = 0.5$, the most likely number of heads is five, followed by four and six. When $\pi = 0.9$, the most likely number of heads is nine, followed by ten and eight.

B.1.2 Maximum Likelihood Estimation

Figure B.1 shows the distribution of X given π. However, typically we can only observe the value of X, and we would like to know π. This is known as statistical inference. For example, suppose we observe five heads (i.e., $X = 5$) on $k = 10$ flips. Looking at Figure B.1, we see that we are likely to observe five heads when $\pi = 0.5$, but not when $\pi = 0.1$ or $\pi = 0.9$. Intuitively, this suggests that 0.5 is a "better estimate" for π than 0.1 or 0.9.

We can understand what makes $\pi = 0.5$ a better estimate by thinking about the number of heads predicted on a future sequence of ten flips by each of the different values of π. Suppose we estimate π to be 0.1. The left panel of Figure B.1 shows that observing five heads on a subsequent experiment of ten flips is unlikely, about 0.00148. That means if we ran 1000 experiments of ten flips, we would likely observe five heads only once or twice. This seems like a somewhat strange prediction, given that we have already observed a result of five heads. The same is true when estimating $\pi = 0.9$.

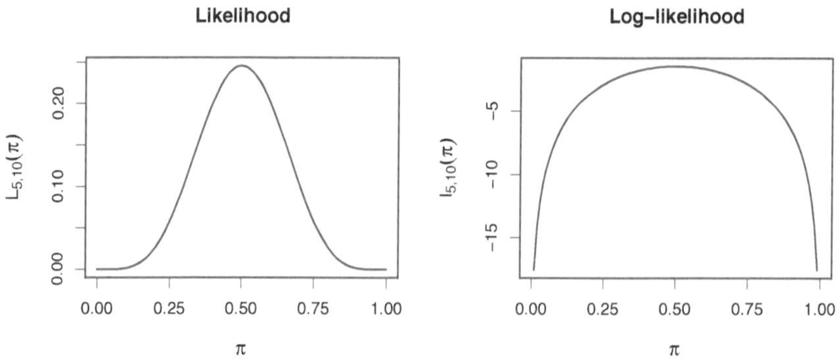

FIGURE B.2: Graphs of the likelihood and log-likelihood for one experiment of ten coin flips.

Now suppose that we estimate π to be 0.5. In this case, Figure B.1 shows that a subsequent observation of five heads is much more likely, about 0.264. This means that a little more than one in every four experiments would result in five heads, and results close to five are also quite likely. This prediction makes more sense. We have just observed five heads, so it seems reasonable to predict that we will observe similar values on future experiments. This is the essence of maximum likelihood estimation.

Maximum likelihood estimation finds the value of π that is most plausible given the results we have seen so far by means of maximizing $\Pr(X = x \mid \pi, k)$. To emphasize that we are treating $\Pr(X = x \mid \pi, k)$ as a function of π, rather than x or k, we write $\Pr(X = x \mid \pi, k)$ as $L_{x,k}(\pi)$. Subscripting by x and k indicates that these values are fixed, because we have already observed the data. For the coin flipping example, the likelihood is

$$L_{5,10}(\pi) = \binom{10}{5} \cdot \pi^5 \cdot (1-\pi)^{10-5} = \binom{10}{5} \cdot \pi^5 \cdot (1-\pi)^5$$

Substituting these values for x and k demonstrates that fixing these values results in a function of a single variable, π. The likelihood is shown in the left panel of Figure B.2.

Comparing Figure B.1 with the left panel of Figure B.2 demonstrates how thinking of $\binom{k}{x} \cdot \pi^x \cdot (1-\pi)^{k-x}$ as a function of x with π fixed, as in the probability $\Pr(X = x \mid \pi, k)$, differs from treating it as a function of π with x fixed, as in the likelihood $L_{x,k}(\pi)$. For example, when viewed as a probability, the function has a domain of $\{0, 1, \ldots, 10\}$. When viewed as a likelihood, the function has a domain of $(0, 1)$.

In most applications, we deal with the natural logarithm of the likelihood function, or log-likelihood, rather than the likelihood itself, because the log-likelihood is easier to handle mathematically. For the coin flipping example, the log-likelihood is

$$l_{5,10}(\pi) = \log\left(L_{5,10}(\pi)\right) = \log\binom{10}{5} + 5 \cdot \log(\pi) + 5 \cdot \log(1-\pi).$$

The graph of this function is shown in the right panel of Figure B.2.

The expression for the log-likelihood demonstrates a useful property of the logarithm: it transforms products into sums (see Appendix A). Sums tend to be easier to deal with than products in derivatives. It also turns out that the likelihood and log-likelihood will always have their maximums at the same value, so taking the logarithm doesn't change the result.

For simple problems, like the coin flipping example, we can find the maximum of the likelihood analytically using calculus. To do this, we start by computing the derivative of the likelihood or log-likelihood. We then find the parameter value for which the derivative is zero. In the coin flipping example, we can solve either

$$\frac{d}{d\pi} L_{5,10}(\pi) = \frac{d}{d\pi}\left[\binom{10}{5} \cdot \pi^5 \cdot (1-\pi)^5\right] = 0$$

or

$$\frac{d}{d\pi} l_{5,10}(\pi) = \frac{d}{d\pi}\left[\log\binom{10}{5} + 5 \cdot \log(\pi) + 5 \cdot \log(1-\pi)\right] = 0.$$

The second is easier to solve, because the sum rule is easier to apply than the product rule.

We can take the derivative as follows. First, we apply the sum rule (for all rules used here see Appendix A) to obtain

$$\begin{aligned}
\frac{d}{d\pi} l_{5,10}(\pi) &= \frac{d}{d\pi}\left[\log\binom{10}{5} + 5 \cdot \log(\pi) + 5 \cdot \log(1-\pi)\right] \\
&= \frac{d}{d\pi}\left[\log\binom{10}{5}\right] + \frac{d}{d\pi}\left[5 \cdot \log(\pi)\right] + \frac{d}{d\pi}\left[5 \cdot \log(1-\pi)\right] \\
&= \frac{d}{d\pi}\left[5 \cdot \log(\pi)\right] + \frac{d}{d\pi}\left[5 \cdot \log(1-\pi)\right],
\end{aligned}$$

since $\log\binom{10}{5}$ is a constant and thus has a derivative of zero. We can then apply the rule for constant multipliers to move the five out of the derivatives in the remaining two terms, yielding

$$\frac{d}{d\pi} l_{5,10}(\pi) = 5 \cdot \frac{d}{d\pi}\left[\log(\pi)\right] + 5 \cdot \frac{d}{d\pi}\left[\log(1-\pi)\right].$$

The derivative of $\log(\pi)$ is $1/\pi$. Moreover, applying the log and chain rule, we find that

$$\frac{d}{d\pi}\left[\log(1-\pi)\right] = -\frac{1}{1-\pi}.$$

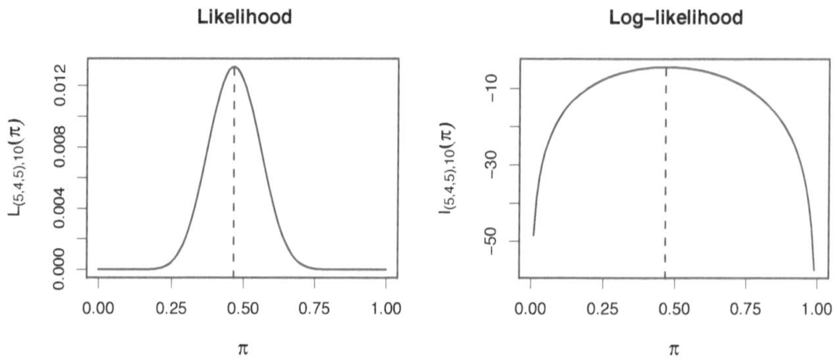

FIGURE B.3: Graphs of the likelihood and log-likelihood for three experiments of ten coin flips.

Substituting these derivatives back into the equation yields

$$\frac{d}{d\pi} l_{5,10}(\pi) = 5 \cdot \frac{1}{\pi} + 5 \cdot \left(-\frac{1}{1-\pi} \right) = 5 \cdot \frac{1}{\pi} - 5 \cdot \frac{1}{1-\pi}.$$

Thus, the maximum likelihood estimate $\hat{\pi}$ is the solution to

$$5 \cdot \frac{1}{\pi} - 5 \cdot \frac{1}{1-\pi} = 0.$$

Solving for π by expanding and rearranging, we get that

$$\begin{aligned} \frac{5}{\pi} &= \frac{5}{1-\pi}, \\ 5 \cdot (1-\pi) &= 5 \cdot \pi, \\ 5 - 5 \cdot \pi &= 5 \cdot \pi, \\ 5 &= 10 \cdot \pi, \end{aligned}$$

which is solved by $\hat{\pi} = \frac{5}{10} = 0.5$.

This means that the data we observed in our first experiment, five heads on ten flips, is perfectly in line with the coin being fair. Still, we would have more confidence in our estimate if we could combine the information from several experiments. This is shown in the next section.

B.1.3 Likelihood for Multiple Observations

Suppose we ask P of our friends to help us collect data by each throwing the coin ten times. This means that we observe P independent experiments of ten (or generally speaking k) flips, $\boldsymbol{x} = (x_1, \ldots x_P)$. Because the observations are independent, the joint probability of the P flips is the product of their individual probabilities,

$$\begin{aligned}\Pr(\boldsymbol{X} = \boldsymbol{x} \mid \pi, k) &= \Pr(X_1 = x_1, \ldots, X_P = x_P \mid \pi, k) \\ &= \prod_{p=1}^{P} \binom{k}{x_p} \cdot \pi^{x_p} \cdot (1-\pi)^{k-x_p}.\end{aligned}$$

The likelihood of π, $L_{\boldsymbol{x},k}(\pi)$, is simply $\Pr(\boldsymbol{X} = \boldsymbol{x} \mid \pi, k)$ expressed as a function of π. For example, suppose $P = 3$ of our friends have helped us run three independent experiments of $k = 10$ coin flips each. In the first experiment we have observed $X_1 = 5$ heads, as above. On the second experiment, we have observed $X_2 = 4$ heads, and on the third $X_3 = 5$ heads again. Then, the joint likelihood of π is

$$\begin{aligned}L_{\boldsymbol{x},k}(\pi) &= \Pr(X_1 = 5, X_2 = 4, X_3 = 5 \mid \pi, k = 10) \\ &= \left[\binom{10}{5} \cdot \pi^5 \cdot (1-\pi)^5\right] \times \left[\binom{10}{4} \cdot \pi^4 \cdot (1-\pi)^6\right] \times \left[\binom{10}{5} \cdot \pi^5 \cdot (1-\pi)^5\right].\end{aligned}$$

This is shown along with the corresponding log-likelihood in Figure B.3.

Just as in the case with a single observation, it is easier to maximize the log-likelihood than it is to maximize the likelihood itself. The log-likelihood is

$$\begin{aligned}l_{\boldsymbol{x},k}(\pi) &= \log\left[\prod_{p=1}^{P} \binom{k}{x_p} \cdot \pi^{x_p} \cdot (1-\pi)^{k-x_p}\right] \\ &= \sum_{p=1}^{P} \log\left[\binom{k}{x_p} \cdot \pi^{x_p} \cdot (1-\pi)^{k-x_p}\right] \\ &= \sum_{p=1}^{P} \left[\log\binom{k}{x_p} + x_p \log(\pi) + (k - x_p)\log(1-\pi)\right] \\ &= \left[\sum_{p=1}^{P} \log\binom{k}{x_p}\right] + \left(\sum_{p=1}^{P} x_p\right) \cdot \log(\pi) + \left[\sum_{p=1}^{P} (k - x_p)\right] \cdot \log(1-\pi) \\ &= \left[\sum_{p=1}^{P} \log\binom{k}{x_p}\right] + \left(\sum_{p=1}^{P} x_p\right) \cdot \log(\pi) + \left(P \cdot k - \sum_{p=1}^{P} x_p\right) \cdot \log(1-\pi).\end{aligned}$$

We can find the maximum of the log-likelihood, and consequently the maximum likelihood estimate, by finding the zeros of its derivative with respect to π. This is

$$\begin{aligned}\frac{d}{d\pi} l_{\boldsymbol{x},k}(\pi) = \frac{d}{d\pi}\Bigg\{&\sum_{p=1}^{P}\left[\log\binom{k}{x_p}\right] + \left(\sum_{p=1}^{P} x_p\right) \cdot \log(\pi) \\ &+ \left(P \cdot k - \sum_{p=1}^{P} x_p\right) \cdot \log(1-\pi)\Bigg\}.\end{aligned}$$

We can simplify this expression by applying the sum rule, resulting in

$$\begin{aligned}\frac{d}{d\pi} l_{\boldsymbol{x},k}(\pi) = \frac{d}{d\pi}\left\{\sum_{p=1}^{P}\left[\log\binom{k}{x_p}\right]\right\} + \frac{d}{d\pi}\left[\left(\sum_{p=1}^{P} x_p\right)\cdot\log(\pi)\right] \\ + \frac{d}{d\pi}\left[\left(P\cdot k - \sum_{p=1}^{P} x_p\right)\cdot\log(1-\pi)\right].\end{aligned}$$

The first term in the sum does not depend on π, so its derivative is zero. Moreover, since $\sum_{p=1}^{P} x_p$ and $P \cdot k - \sum_{p=1}^{P} x_p$ are constant multipliers,

$$\begin{aligned}\frac{d}{d\pi} l_{\boldsymbol{x},k}(\pi) &= \left(\sum_{p=1}^{P} x_p\right)\cdot\frac{d}{d\pi}\log(\pi) + \left(P\cdot k - \sum_{p=1}^{P} x_p\right)\cdot\frac{d}{d\pi}\log(1-\pi) \\ &= \left(\sum_{p=1}^{P} x_p\right)\cdot\frac{1}{\pi} - \left(P\cdot k - \sum_{p=1}^{P} x_p\right)\cdot\frac{1}{1-\pi}.\end{aligned}$$

Setting $\frac{d}{d\pi} l_{\boldsymbol{x},k}(\pi) = 0$ results in the equation

$$\left(\sum_{p=1}^{P} x_p\right)\cdot\frac{1}{\pi} + \left(P\cdot k - \sum_{p=1}^{P} x_p\right)\cdot\frac{1}{1-\pi} = 0,$$

which is solved (following the same steps as above) by

$$\hat{\pi} = \frac{\sum_{p=1}^{P} x_p}{P\cdot k}.$$

This means that the best estimate for π is the average percentage of heads among all flips, which for this example gives $\frac{5+4+5}{3\cdot 10} = \frac{14}{30}$ or about 0.467. This value for the maximum likelihood estimate $\hat{\pi}$ is indicated as a dashed line in Figure B.3. Note also that the joint likelihood of the three experiments depicted in Figure B.3 (left) is tighter than that of a single experiment in Figure B.2 (left), because it contains more information.

For our example, this means that the data from the three experiments (even if we did not observe five heads in every try) should leave us even more confident that a probability of 0.5, which we expect from a fair coin, is among the most plausible values. Were we to repeat the experiment again and again, the estimate would fluctuate randomly, but on average reflect the true probability with less and less uncertainty. Confidence intervals reflecting the respective degree of uncertainty can be derived based on the asymptotic normality of maximum likelihood estimates.

B.1.4 Bayesian Inference

Bayes' rule relates the conditional probabilities $\Pr(A \mid B)$ and $\Pr(B \mid A)$ for events A and B. It states that

$$\Pr(B \mid A) = \frac{\Pr(A \mid B) \cdot \Pr(B)}{\Pr(A)}.$$

Bayesian inference methods are based on Bayes' rule. Rather than maximizing the likelihood, Bayesian methods compute the posterior distribution. This is the distribution of the unknown parameter given the observed data.

B.1.4.1 Bayes' Rule by Example

Bayes' rule is easiest to understand by example. Suppose a patient tests positive on a diagnostic test for cancer. Diagnostic tests are not perfect. Sometimes they will give a positive result for patients without cancer. We would like to know the probability that a patient has cancer given that the diagnostic test was positive.

Let A be the event "tests positive for cancer" and B be the event "has cancer". Then, Bayes' rule states that

$$\Pr(\text{has cancer} \mid \text{tests positive}) = \frac{\Pr(\text{tests positive} \mid \text{has cancer}) \cdot \Pr(\text{has cancer})}{\Pr(\text{tests positive})}.$$

Suppose that the incidence rate for cancer is 10%, which means $\Pr(\text{has cancer}) = 0.1$ and $\Pr(\text{no cancer}) = 0.9$. Moreover, suppose that, among people who have cancer, the test comes back positive 95% of the time. This means that $\Pr(\text{tests positive} \mid \text{has cancer}) = 0.95$. Finally, suppose that the test comes back positive for 30% of patients without cancer, meaning that $\Pr(\text{tests positive} \mid \text{no cancer}) = 0.3$.

We can compute the denominator $\Pr(\text{tests positive})$ – the total probability that a patient tests positive for cancer – by noting that a patient can test positive for cancer when they have cancer or when they do not. The total probability that a patient tests positive for cancer is the sum of these probabilities, $\Pr(\text{tests positive, has cancer}) + \Pr(\text{tests positive, no cancer})$. We can compute these probabilities by factoring the joint probability into a product of conditional and marginal probabilities, i.e.,

$$\begin{aligned}
\Pr(\text{tests positive, has cancer}) &= \Pr(\text{tests positive} \mid \text{has cancer}) \cdot \Pr(\text{has cancer}) \\
&= 0.95 \times 0.1 \\
&= 0.095
\end{aligned}$$

and

$$\begin{aligned}
\Pr(\text{tests positive, no cancer}) &= \Pr(\text{tests positive} \mid \text{no cancer}) \cdot \Pr(\text{no cancer}) \\
&= 0.3 \times 0.9 \\
&= 0.27.
\end{aligned}$$

Thus,

$$\begin{aligned}\Pr(\text{tests positive}) &= \Pr(\text{tests positive, has cancer}) \\ &\quad + \Pr(\text{tests positive, no cancer}) \\ &= 0.095 + 0.27 \\ &= 0.365.\end{aligned}$$

We can then compute the probability that a patient has cancer given a positive test result by substituting Pr(tests positive | has cancer), Pr(has cancer), and Pr(tests positive). The result is

$$\begin{aligned}\Pr(\text{has cancer} \mid \text{tests positive}) &= \frac{\Pr(\text{tests positive} \mid \text{has cancer}) \cdot \Pr(\text{has cancer})}{\Pr(\text{tests positive})} \\ &= \frac{0.95 \times 0.1}{0.365} \\ &\approx 0.26.\end{aligned}$$

This means that there is only a one in four chance that the patient has cancer, despite the positive test. You can repeat the computation for different values and will see that this effect – that the probability that a patient has the disease given a positive test result is surprisingly low, despite the high accuracy of the test – is particularly striking for diseases with a low incidence rate.

We can apply the logic of the cancer example to statistical inference generally. In inference problems, we observe data which we would like to use to draw conclusions about an unknown parameter. In the example, the data is the result of the diagnostic test and the unknown parameter is a dichotomous variable indicating whether the patient has cancer. The probability Pr(tests positive | has cancer) is known as the likelihood. The likelihood in Bayesian inference differs conceptually from the likelihood in maximum likelihood. In Bayesian inference, the likelihood is treated as a probability rather than a function of the unknown parameter.

The probability Pr(has cancer) is called the prior distribution, or just the prior. In the cancer example, the prior distribution specifies the incidence rate of cancer. More generally, the prior distribution encodes any information that is available about the unknown parameter before administering a test or experiment. In the coin flipping example, the prior distribution $f(\pi)$ would encode any information about π that is available before the first coin flip.

The probability Pr(has cancer | tests positive) is called the posterior distribution. In the cancer example, the posterior distribution specifies the probability that the patient has cancer given that their diagnostic test was positive. In the coin flipping example, the posterior distribution $f(\pi \mid X = x)$ specifies the density of each value of π after observing the coin flips. The posterior distribution is related to the prior distribution through Bayes' rule. This means that

$$f(\pi \mid X = x) = \frac{\Pr(X = x \mid \pi) \cdot f(\pi)}{\Pr(X = x)}.$$

The denominator $\Pr(X = x)$ is the total probability that $X = x$. The total probability aggregates the probability that $X = x$ for every possible value of π. It is completely determined by the likelihood and prior. In the coin flipping example, it is given by the integral

$$\Pr(X = x) = \int_0^1 \Pr(X = x \mid \pi) \cdot f(\pi)\, d\pi.$$

In some cases, it is possible to compute $\Pr(X = x)$ analytically. However, for the Rasch model, we will not be able to compute the denominator of Bayes' rule analytically. Instead, we will apply Markov chain Monte Carlo (MCMC) methods to sample the posterior.

B.1.4.2 Coin Flipping with a Uniform Prior

The coin flipping example is useful for demonstrating how the prior distribution affects inference. Recall that the likelihood for the coin flipping example is

$$\Pr(X = x \mid \pi, k) = \binom{k}{x} \cdot \pi^x \cdot (1-\pi)^{k-x}.$$

When $x = 5$ and $k = 10$, this corresponds to

$$\Pr(X = 5 \mid \pi, k = 10) = \binom{10}{5} \cdot \pi^5 \cdot (1-\pi)^5.$$

To employ Bayesian inference, we need to pick a prior $f(\pi)$ for π. A simple choice of prior is the uniform distribution on the interval from zero to one, which assigns a density of one to every value of π between zero and one, i.e.,

$$f(\pi) = 1.$$

Since every value of π is equally likely, the uniform distribution is often called an uninformative prior distribution.

In this example, we can compute $\Pr(X = x)$ analytically for the uniform prior distribution. When $x = 5$ and $k = 10$,

$$\begin{aligned}
\Pr(X = 5) &= \int_0^1 \Pr(X = 5 \mid \pi, k = 10) \cdot f(\pi)\, d\pi \\
&= \int_0^1 \binom{10}{5} \cdot \pi^5 \cdot (1-\pi)^5 \cdot 1\, d\pi \\
&= \int_0^1 \binom{10}{5} \cdot \pi^5 \cdot (1-\pi)^5\, d\pi.
\end{aligned}$$

Because $\binom{10}{5}$ is a constant, we can pull it out of the integral giving

$$\Pr(X = 5) = \binom{10}{5} \cdot \int_0^1 \pi^5 \cdot (1-\pi)^5\, d\pi.$$

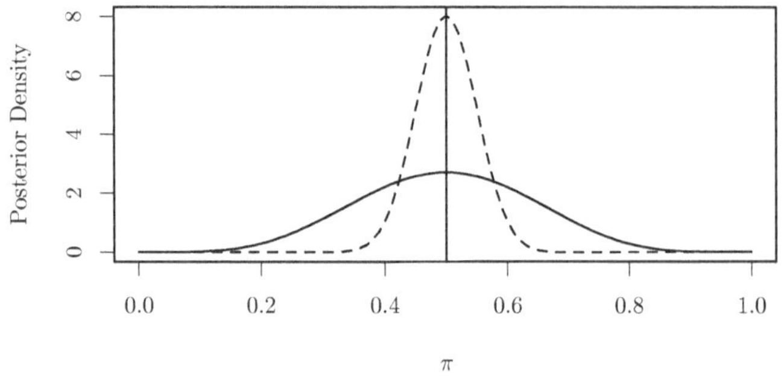

FIGURE B.4: The solid line shows the posterior distribution of π when $X = 5$ and $k = 10$, $f(\pi \mid X = 5, k = 10)$. The dashed line shows the posterior distribution of π when $X = 50$ and $k = 100$, $f(\pi \mid X = 50, k = 100)$. The vertical line shows the maximum likelihood estimate, 0.5.

This leaves us with the integral

$$\int_0^1 \pi^5 \cdot (1-\pi)^5 \, d\pi.$$

This integral has a special form, namely that of the so called beta function, which is generally defined as:

$$B(a,b) = \int_0^1 \pi^{a-1} \cdot (1-\pi)^{b-1} \, d\pi$$

Thus we can abbreviate the integral $\int_0^1 \pi^5 \cdot (1-\pi)^5 \, d\pi$ as B(6, 6), and in summary we obtain:

$$\Pr(X = 5) = \binom{10}{5} \cdot B(6,6).$$

Using the same derivation, we obtain the following formula for arbitrary x and k:

$$\Pr(X = x) = \binom{k}{x} \cdot B(x+1, k-x+1)$$

We can combine $\Pr(X = x \mid \pi, k)$, $f(\pi)$ and $\Pr(X = x)$ to obtain the posterior distribution. When $x = 5$ and $k = 10$, the posterior distribution is

$$\begin{aligned} f(\pi \mid X = 5, k = 10) &= \frac{\Pr(X = 5 \mid \pi, k = 10) \cdot f(\pi)}{\Pr(X = 5)} \\ &= \frac{\binom{10}{5} \cdot \pi^5 \cdot (1-\pi)^5 \cdot (1)}{\binom{10}{5} \cdot B(6,6)} \\ &= \frac{\pi^5 \cdot (1-\pi)^5}{B(6,6)}, \end{aligned}$$

since the $\binom{10}{5}$ is in the numerator and denominator.

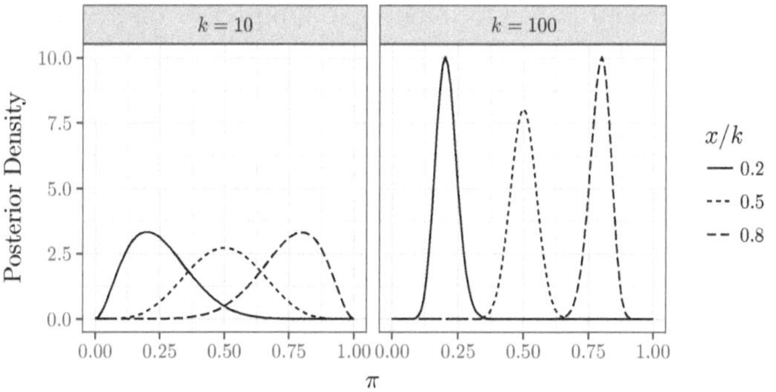

FIGURE B.5: The beta distribution for different proportions of heads x/k. In each plot, the solid line corresponds to $x/k = 0.2$, the densely dashed line to $x/k = 0.5$ and the dashed line to $x/k = 0.8$.

The resulting density for $f(\pi \mid X = 5, k = 10)$ also has a special form, namely that of the probability density function of the so called beta distribution. Generally, the probability density function f_{Beta} of the beta distribution, $\text{Beta}(a, b)$, is

$$f_{\text{Beta}}(\pi, a, b) = \frac{\pi^{a-1} \cdot \pi^{b-1}}{B(a, b)},$$

where $B(a, b)$ is the beta function and a and b are shape parameters. Note for clarity that this notation distinguishes between the beta *function*, $B(a, b)$, from above and the beta *distribution*, $\text{Beta}(a, b)$, which we have just introduced.

In our example, the posterior $\frac{\pi^5 \cdot (1-\pi)^5}{B(6,6)}$ corresponds to a beta distribution with $a = 6$ and $b = 6$, abbreviated by $\text{Beta}(6, 6)$.

The graph of the posterior $\text{Beta}(6, 6)$ distribution is shown in Figure B.4 as the solid line, along with the maximum likelihood estimate (the vertical line). This plot shows that the posterior distribution, in this particular example, is symmetric and centered at the maximum likelihood estimate, 0.5. It also shows that the posterior distribution is more concentrated around the estimate when the number of flips is higher (dashed line).

Highest posterior density (HPD) intervals are a common way to summarize uncertainty about the true parameter value. A 95% HPD interval is the shortest interval containing 95% of the area under the posterior density distribution. For unimodal distributions, like $f(\pi \mid X = 5, k = 10)$, we can compute the endpoints of the 95% HPD interval by computing the 0.025 and 0.975 quantiles of the posterior distribution. These are about 0.23 and 0.76 in this case. Thus, there is still considerable uncertainty about the location of π after learning the result of the ten coin flips.

For arbitrary x and k, the formula for the posterior distribution is

$$f(\pi \mid X = x, k) = \frac{\pi^x \cdot (1-\pi)^{k-x}}{B(x+1, k-x+1)}.$$

This is the density function of a Beta$(x+1, k-x+1)$-distributed random variable, i.e., a beta-distributed random variable with shape parameters $a = x+1$ and $b = k-x+1$. The effect of x and k on the posterior distribution shape are shown in Figure B.5. We can see that the proportion of heads x/k determines the location of the posterior distribution, with larger ratios corresponding to posterior distributions located closer to one and smaller ratios corresponding to posterior distributions located closer to zero. The spread of the posterior distributions is determined by the number of flips k. Larger values of k correspond to tighter posterior distributions. This is sensible, since each flip provides additional information about the probability of heads.

B.1.4.3 Informative Priors and Beta-Binomial Model

Suppose that, before running the current experiment, we had already run the experiment once and observed five heads in ten coin flips. After analyzing the experiment using a uniform prior, we found that the posterior distribution for π is a Beta$(6, 6)$ distribution. The previous experiment offers information about π that is available prior to running the current experiment. We can incorporate this prior information into our analysis by using the posterior of the previous experiment as the prior for the current experiment. The Beta$(6, 6)$ prior is an informative prior distribution for our current experiment.

We can compute the posterior distribution for the Beta$(6, 6)$ prior distribution as follows. Bayes' rule states that

$$f(\pi \mid X = 5, k = 10) = \frac{\Pr(X = 5 \mid \pi, k = 10) \cdot f(\pi)}{\Pr(X = 5)}.$$

When the prior $f(\pi)$ is the Beta$(6, 6)$ distribution, the numerator is

$$\begin{aligned}
\Pr(X = 5 \mid \pi, k = 10) \cdot f(\pi) &= \left[\binom{10}{5} \cdot \pi^5 \cdot (1-\pi)^5\right] \cdot \left[\frac{\pi^5 \cdot (1-\pi)^5}{B(6,6)}\right] \\
&= \frac{1}{B(6,6)} \cdot \binom{10}{5} \pi^{10} \cdot (1-\pi)^{10}.
\end{aligned}$$

The denominator is

$$\begin{aligned}
\Pr(X = 5) &= \int_0^1 \left[\binom{10}{5} \cdot \pi^5 \cdot (1-\pi)^5\right] \cdot \left[\frac{\pi^5 \cdot (1-\pi)^5}{B(6,6)}\right] d\pi \\
&= \frac{1}{B(6,6)} \cdot \binom{10}{5} \cdot \int_0^1 \pi^{10} \cdot (1-\pi)^{10} \, d\pi.
\end{aligned}$$

Recognizing that

$$\int_0^1 \pi^{10} \cdot (1-\pi)^{10} = B(11, 11),$$

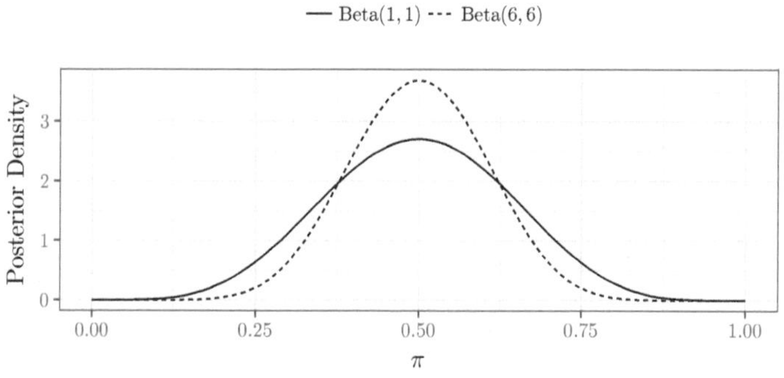

FIGURE B.6: Posterior distributions after observing five heads in ten flips for Beta(1, 1) and Beta(6, 6) prior distributions.

we obtain

$$\Pr(X = 5) = \frac{1}{B(6,6)} \cdot \binom{10}{5} \cdot B(11, 11).$$

Combining these results, we obtain

$$\begin{aligned} f(\pi \mid X = 5, k = 10) &= \frac{\Pr(X = 5 \mid \pi, k = 10) \cdot f(\pi)}{\Pr(X = 5)} \\ &= \frac{\frac{1}{B(6,6)} \cdot \binom{10}{5} \cdot \pi^{10} \cdot (1-\pi)^{10}}{\frac{1}{B(6,6)} \cdot \binom{10}{5} \cdot B(11,11)} \\ &= \frac{\pi^{10} \cdot (1-\pi)^{10}}{B(11,11)} \end{aligned}$$

after cancellation. Thus, the posterior distribution is a Beta(11, 11) distribution.

In the general case where the prior for π is a Beta(a_0, b_0) distribution, the posterior distribution is a Beta($x + a_0, k - x + b_0$) distribution. Intuitively, a_0 and b_0 can be thought of as the prior sample size. When the prior number of heads a_0 is larger than the prior number of tails b_0, the prior favors values of π greater than 0.5. This makes sense, since π is the probability of flipping heads. When the prior number of heads a_0 is smaller than the prior number of tails b_0, the prior favors values of π smaller than 0.5. Finally, when $a_0 = b_0$, the prior is symmetric about 0.5. The spread is determined by the prior sample size $a_0 + b_0$. The larger $a_0 + b_0$, the more concentrated the prior.

Figure B.6 demonstrates the effect of the prior on the posterior. It plots the posterior distributions resulting from an uninformative Beta(1, 1) prior (solid line) and an informative Beta(6, 6) prior (dashed line). We see that the posterior distribution resulting from a Beta(6, 6) prior is more concentrated at 0.5 than the posterior resulting from a Beta(1, 1) prior. This is due to the fact that the posterior under the Beta(6, 6) essentially combines the information

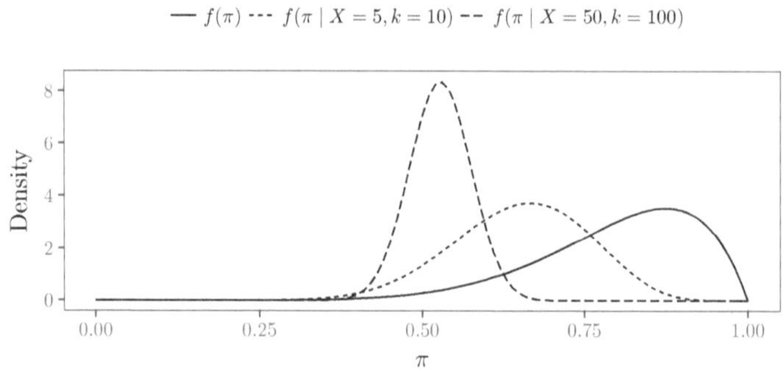

FIGURE B.7: Beta(8, 2) prior distribution (solid line) and its corresponding posterior distributions (dashed lines) after observing five heads in ten flips and 50 heads in 100 flips.

of two experiments of ten coin flips. It also suggests that the prior distribution can play an important role in shaping the posterior distribution.

Figure B.7 demonstrates how the prior affects the shape of the posterior distribution. The solid line in Figure B.7 shows the Beta(8, 2) prior distribution. This prior favors large values of π, in particular those above 0.6. Figure B.7 also shows the posterior distributions after observing five heads in ten flips, $f(\pi \mid X = 5, k = 10)$ and 50 heads in 100 flips, $f(\pi \mid X = 50, k = 100)$. Both of these posterior distributions favor values of π above 0.5, demonstrating how the posterior distribution balances the prior against the observed data. In this case, the prior distribution indicates that π should be somewhere above 0.6, most likely around 0.8, and the data indicates that π should be near 0.5.

Comparing the shape of $f(\pi \mid X = 5, k = 10)$ to that of $f(\pi \mid X = 50, k = 100)$ demonstrates how the posterior distribution weights the prior and likelihood according to their respective sample sizes. When the prior sample size $a_0 + b_0$ is larger than the observed sample size k, the posterior will weight the prior more than the likelihood and stay closer to the prior on the right. When the prior sample size is smaller than the observed sample size, the posterior distribution will weight the prior less than the likelihood and move further away from the prior towards 0.5.

B.2 Statistical Testing

This section demonstrates how to make statistical decisions using rejection regions and p-values. It forms the basis for the tests employed in this book.

	No cat	Cat	Total
No dog	70	30	100
Dog	30	70	100
Total	100	100	200

TABLE B.1: Results of an example survey about cat and dog ownership.

B.2.1 Tests Based on the χ^2 Distribution

B.2.1.1 χ^2 Test for Independence

Suppose you are interested in whether there is a relationship between having a cat and having a dog. To find out, you draw a random sample of 200 people and ask them whether they have a cat, a dog or both. The results are shown in Table B.1. We can test whether the two variables – having a cat and having a dog – are independent using a χ^2 test for independence. This test compares the observed numbers of people with dogs and cats in Table B.1 to the numbers that we would expect to see if having a cat and having a dog were independent.

We can compute the expected numbers of people with cats and dogs using the "Total" rows and columns. The total columns tell us that half of our sample owns cats. Thus, we estimate the marginal probability of owning a cat to be $\Pr(\text{cat}) = \frac{100}{200} = 0.5$. Similarly, the total rows tell us that half of our sample owns dogs, so we estimate the marginal probability of owning a dog to be $\Pr(\text{dog}) = \frac{100}{200} = 0.5$. When having a cat is statistically independent of having a dog, we expect the joint probability of having a cat and a dog to be

$$\Pr(\text{cat}, \text{dog}) = \Pr(\text{cat}) \times \Pr(\text{dog}) = 0.5 \times 0.5 = 0.25.$$

This means that, from our 200 person sample, we expect $200 \times \Pr(\text{cat}, \text{dog}) = 200 \times 0.25 = 50$ people to have both a dog and a cat. Continuing this argument, we see that

$$\Pr(\text{cat}, \text{no dog}) = \Pr(\text{cat}) \times [1 - \Pr(\text{dog})] = 0.5 \times 0.5 = 0.25,$$

so the expected number of people who have a cat but not a dog is also 50. Repeating these calculations, we find that in this simple example the expected number of people having a dog but not a cat is also 50, as is the number of people without a dog or cat.

Now that we know the expected frequencies, we can compare them to the observed frequencies to compute the test statistic for the χ^2 test for independence. Let O_c be the observed frequency in a single cell c of Table B.1 and let E_c be its corresponding expected frequency. Then, the χ^2 test statistic is

$$T = \sum_c \frac{(O_c - E_c)^2}{E_c}, \tag{B.1}$$

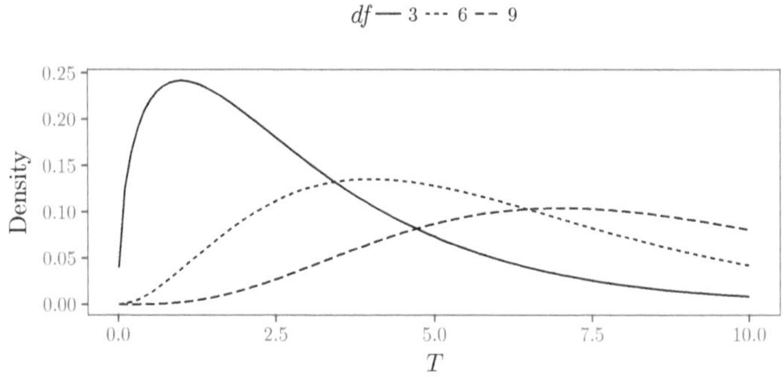

FIGURE B.8: The χ^2 distribution for different degrees of freedom.

where the sum runs over all cells of Table B.1. Note that this statistic can only take positive values. In this simple example, $O_c - E_c$ is $70 - 50 = 20$ for people with cats and dogs and people without a dog or cat and $30 - 50 = -20$ for people with cats but not dogs or dogs but not cats. Squaring these values and dividing each by its expected frequency gives

$$\frac{(70-50)^2}{50} + \frac{(30-50)^2}{50} + \frac{(30-50)^2}{50} + \frac{(70-50)^2}{50} = 32.$$

We can use the test statistic to evaluate whether our sample is consistent with the null hypothesis that having a cat and having a dog are independent. Stronger relationships between the variables will tend to result in larger deviations between the observed and expected frequencies. Since the test statistic increases with these deviations, it can be thought of as an index of the evidence against independence provided by the sample.

Though we tend to observe large values of the test statistic when there actually is a relationship between having a cat and having a dog, we can still observe values different from zero when there is not. Most random samples will not contain exactly the numbers of cat and dog owners expected under the null hypothesis, resulting in positive values of the test statistic. The distribution of the test statistic across random samples of people is known as the sampling distribution of the test statistic under the null hypothesis. This distribution tells us which values of the test statistic are likely to occur by chance when the null hypothesis is true and which are not. Observing a value of T that is unlikely under the null hypothesis provides evidence that the null hypothesis is not true.

Unsurprisingly, the sampling distribution under the null hypothesis for the χ^2 test of independence is (approximately) a χ^2 distribution. The χ^2 distribution has a single parameter, the number of the degrees of freedom, which determines its shape. The χ^2 distribution is shown for a number of different degrees of freedom *df* in Figure B.8. It demonstrates that increasing

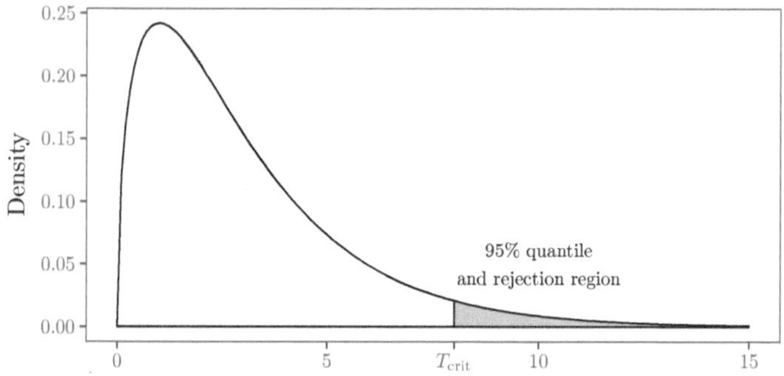

FIGURE B.9: Illustration of the rejection region in the χ^2 test of independence. The gray area illustrates a 5% rejection region for the χ^2 distribution with three degrees of freedom.

the degrees of freedom moves the center of the distribution to the right and increases the spread of values that we could reasonably observe.

The degrees of freedom for the χ^2 test of independence are the number of cells in the table minus one. For our example, the degrees of freedom $df = 3$. We can use the sampling distribution to define a threshold T_{crit}, known as the critical value, for the observed values of the test statistic. When $T > T_{\text{crit}}$, we reject the null hypothesis that having a cat and having a dog are independent. The region where $T > T_{\text{crit}}$ is known as the rejection region.

We can never be 100% sure of a rejection of the null hypothesis. Randomly sampling from a population where having a cat and having a dog are independent will occasionally produce a sample where all or most people with cats also have dogs and most people with dogs also have cats, leading to a large value of the test statistic. Thankfully, we can use the sampling distribution to limit the probability of incorrectly rejecting the null hypothesis to some small value α – the significance level – typically chosen to be 0.05 or 0.01.

This probability defines the critical value. Suppose we want to ensure that no more than 5% of rejections of the null hypothesis are incorrect. Then, we set the critical value of the test statistic, T_{crit}, to be the 95th percentile of the χ^2 distribution. This ensures that 95% of samples where the two variables are independent will produce a test statistic that is below T_{crit}. This is illustrated by Figure B.9. More generally, when we want to ensure that no more than $100 \times \alpha\%$ of rejections of the null hypothesis are incorrect, we set T_{crit} to be the $100 \times (1 - \alpha)$ percentile of the χ^2 distribution.

Rejecting the null hypothesis when it is true is known as a Type I error. For large samples, the rate of Type I errors will be equal to the significance level α. We can reduce the Type I error by lowering the significance level α, e.g., to 0.01. The cost of this change is an increase in the number of violations of the null hypothesis that go unnoticed (Type II errors). In practice, 0.05 and 0.01

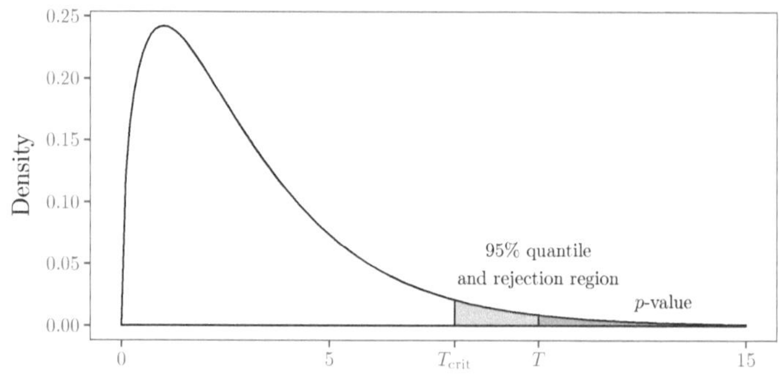

FIGURE B.10: Illustration of the p-value in the χ^2 test of independence indicated by the dark gray area. The light gray area illustrates a 5% rejection region for the χ^2 distribution with three degrees of freedom.

are the most common significance levels, because they provide a good balance between Type I and Type II error rates.

Another way to decide whether the null hypothesis should be rejected is the p-value associated with the test statistic T. The p-value is the probability of sampling a test statistic value that is as extreme or more extreme than the one observed, given that the null hypothesis is true. Thus, instead of comparing T with its critical value, we compute the probability of T directly.

The interpretation of the p-value is illustrated by Figure B.10. A very small p-value indicates that the null hypothesis should be rejected. If the p-value is below the significance level (e.g., below 5%), the null hypothesis is rejected. Therefore, the p-value leads to the same decision as the critical value. As demonstrated by Figure B.10, the p-value falls below α in exactly those cases where the test statistic T falls in the rejection region. Most statistical programs only give p-values for the test decision, because it is straightforward to use p-values for any statistical test at any significance level.

B.2.1.2 Goodness-of-Fit and Other χ^2 Tests

The χ^2 test for independence is a special case of the χ^2 goodness-of-fit test. This test can be used to test whether a model, like the Rasch model, provides a reasonable account of observed data. In the χ^2 test of independence, we compare observed counts with the counts expected under independence. In the χ^2 goodness-of-fit test, we compare observed counts to the counts expected under the Rasch model or some other model.

Performing the χ^2 goodness-of-fit test is exactly the same as performing the χ^2 test of independence. First, we compute the test statistic T by computing Equation (B.1) for the cells of a table. For example, the χ^2 goodness-of-fit test for the Rasch model compares the observed and expected frequencies of

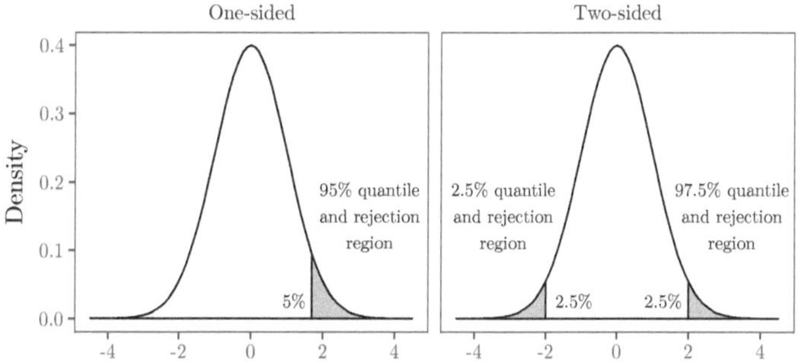

FIGURE B.11: Illustration of rejection region for a standard normal distribution with a one-sided rejection region (left) and a two-sided rejection region (right).

possible response patterns. Thus, each cell of the table is a possible response pattern.

The null hypothesis for the χ^2 goodness-of-fit test asserts that the model of interest, e.g., the Rasch model, generated the observed data. The distribution of the test statistic under the null hypothesis is a χ^2 distribution with *df* degrees of freedom, where *df* is the number of cells in the table minus one. We can then either compare the observed T to the critical value of χ^2 distribution for *df* degrees of freedom at significance level α or compute the p-value associated with T directly. When the observed value of the test statistic is greater than the critical value or when its associated p-value is less than the significance level, we reject the null hypothesis that the model of interest generated the data.

The χ^2 goodness-of-fit test is not the only test based on the χ^2 distribution. Another well-known example is the likelihood ratio test (see Chapter 4). While other tests will differ from the χ^2 goodness-of-fit test in their null hypothesis and test statistic, their overall logic is the same. First, compute the test statistic. Second, determine its degree of freedom. Third, compute either the critical value or p-value. Finally, compare the test statistic to the critical value or the p-value to the significance level. If the test statistic exceeds the critical value or the p-value is lower than the significance level, reject the null hypothesis.

B.2.2 Tests Based on the Normal Distribution

Another distribution that is often used for tests is the (standard) normal distribution. As demonstrated by the left panel of Figure B.11, the normal distribution can be used in the same way as the χ^2 distribution, with very large values (or, conversely, very small values) indicating that the null hypothesis should be rejected. This is called a one-sided test, because we are only interested in whether the test statistic falls in a rejection region located on one side of the normal distribution.

The normal distribution can also be used for two-sided tests, as illustrated by the right panel of Figure B.11. In this case, a very large or a very small value indicates that the null hypothesis should be rejected, as for example in the item-wise Wald test in Chapter 4. In two-sided tests, we halve the significance level α, so that there is a rejection interval of $\alpha/2$ on each side. For example, a two-sided test at the $\alpha = 0.05$ level will have rejection regions with probability 0.025 at either tail of the distribution. Accordingly, the null hypothesis is rejected whenever the test statistic T is small enough to fall in the left part of the rejection region or large enough to fall in the right part.

Statistical software usually provides a two-sided p-value for two-sided tests. The two-sided p-value can be directly compared with the significance value of the test. For example, when $\alpha = 0.05$, we reject the null hypothesis whenever the two-sided p-value is less than 0.05.

C

Answers to the End of Chapter Questions

CONTENTS

C.2 Answers for Chapter 2

1. (a) The difficulty of the item with the solid ICC is 0, because that is the ability with a 0.5 probability of a correct response.
   (b) The item with the dashed ICC is more difficult, because it is to the right of the solid ICC. That means the probability of correctly responding to the dashed item is lower than the probability of correctly responding to the solid item for every ability.
2. A sufficient statistic is a statistic summarizing all of the information that a sample contains about a model parameter. The sufficient statistics of the Rasch model are the row and column sums of the data matrix. The row sums are sufficient statistics for the ability parameters, and the columns sums are sufficient statistics for the difficulty parameters.
3. (a) Local stochastic independence of items does not hold when a person's response to one item depends on their response to another item, for example because the answer to the first item helps solving the second one.

(b) Local stochastic independence of persons does not hold when a person's response to an item depends on another person's response to the item, for example when test takers cheat by copying from each other.

4. The curves do not satisfy specific objectivity, because they cross.
5. The sum scores are only valid as a measure of the persons' abilities if the Rasch model holds. If it is violated, for example by DIF, the sum scores are equally affected by this violation and cannot be used safely for further analyses. As an example, if the aim of the analysis is to investigate ability differences between test takers from different countries, but the test contains items that are easier or harder to solve for test takers from certain countries, this DIF will also affect the sum scores (unless it is perfectly balanced, which is very unlikely in practice, as discussed in Chapter 4). As another example, if the items violate the Rasch model in that they do not have equal slopes, a sum score that weights each item equally does not correctly reflect the underlying ability.

C.3 Answers for Chapter 3

1. The maximum likelihood estimate is the value of θ_p associated with the highest point on the graph. Thus, $\hat{\theta}_p = 1$.
2. (a) Joint maximum likelihood simultaneously finds the maximum of $\boldsymbol{\theta}$ and $\boldsymbol{\beta}$ in the joint likelihood defined by Equation (3.1). Conditional maximum likelihood finds the maximum of $\boldsymbol{\beta}$ conditional on the observed row sums using Equation (3.2). Conditional maximum likelihood is consistent, while joint maximum likelihood is not. Joint maximum likelihood estimates the person and item parameters in a single step, while conditional maximum likelihood first estimates the item parameters and then estimates the person parameters.

   (b) Conditional maximum likelihood finds the maximum of $\boldsymbol{\beta}$ conditional on the observed row sums using Equation (3.2). Marginal maximum likelihood assumes a population distribution for θ_p (typically a normal distribution), averages the joint likelihood over that population distribution, and maximizes the resulting likelihood in $\boldsymbol{\beta}$. Conditional and marginal maximum likelihood are both consistent and both estimate the item parameters before the person parameters.
3. The posterior distribution will contain both the information about θ_p in the prior distribution and the information about θ_p contained

in the data. Because it has more information, the posterior distribution should be more concentrated, or peaked, than the prior. Thus, the prior is the flatter, dotted curve, and the posterior is the more peaked, solid curve.

4. You could simply use the sample median of the MCMC samples in place of the sample mean, which was used on p. 48. You may want to compute the median in cases where the posterior distribution is skewed or when creating distributional summaries.

5. The explicit method for estimating the item parameters can be derived as follows:

   – Under the Rasch model, the probability of a person with ability parameter θ_1 for solving item 1 is

   $$\frac{\exp(\theta_1 - \beta_1)}{1 + \exp(\theta_1 - \beta_1)}$$

   and the probability for not solving item 2 is

   $$\frac{1}{1 + \exp(\theta_1 - \beta_2)}.$$

   By multiplying these probabilities, we get the probability for solving item 1, but not item 2

   $$\frac{\exp(\theta_1 - \beta_1)}{1 + \exp(\theta_1 - \beta_1)} \cdot \frac{1}{1 + \exp(\theta_1 - \beta_2)} = \frac{\exp(\theta_1 - \beta_1)}{(1 + \exp(\theta_1 - \beta_1))(1 + \exp(\theta_1 - \beta_2))}.$$

   By simply switching the indices, we get the probability for solving item 2, but not item 1

   $$\frac{\exp(\theta_1 - \beta_2)}{(1 + \exp(\theta_1 - \beta_1))(1 + \exp(\theta_1 - \beta_2))}.$$

   The ratio of these probabilities is:

   $$\frac{\left(\frac{\exp(\theta_1 - \beta_1)}{(1+\exp(\theta_1 - \beta_1))(1+\exp(\theta_1 - \beta_2))}\right)}{\left(\frac{\exp(\theta_1 - \beta_2)}{(1+\exp(\theta_1 - \beta_1))(1+\exp(\theta_1 - \beta_2))}\right)}.$$

   The terms $(1 + \exp(\theta_1 - \beta_1))(1 + \exp(\theta_1 - \beta_2))$ are in both the numerator and denominator. Therefore they can be reduced, leading to

   $$\frac{\exp(\theta_1 - \beta_1)}{\exp(\theta_1 - \beta_2)}.$$

Using well-known calculation rules for the exponential function (cf. Appendix A), this further simplifies to

$$\exp(\theta_1 - \beta_1 - \theta_1 + \beta_2) = \exp(\beta_2 - \beta_1).$$

- Again using calculation rules for the exponential function on the last expression of the previous step, we get

$$\exp(\beta_2 - \beta_1) = \frac{\exp(\beta_2)}{\exp(\beta_1)} = \frac{\exp(-\beta_1)}{\exp(-\beta_2)} = \frac{\epsilon_1}{\epsilon_2}.$$

The last term obviously does not depend on the ability parameter and is therefore the same for all test takers.

- Plugging in the definitions for the ϵ and using basic calculation rules, we get

$$1 = \prod_i \epsilon_i = \prod_i \exp\left(-\beta_i\right) = \exp\left(-(\sum_i \beta_i)\right).$$

Applying the natural logarithm to the right-hand side and the left-hand side of the previous equation (cf. Appendix A), we get

$$0 = -(\sum_i \beta_i).$$

By multiplying with -1, we get

$$\sum_i \beta_i = 0.$$

- As we have seen, the ratio of the two probabilities for solving item 1, but not 2, and solving item 2, but not 1, is

$$\frac{\epsilon_1}{\epsilon_2}.$$

Since this holds for all test takers, we can approximate the ratio of the two probabilities by the ratio of the two frequencies

$$\frac{\epsilon_1}{\epsilon_2} \approx \frac{p_{12}}{p_{21}},$$

where p_{12} is the number of respondents who solve item 1 but not item 2.

- We now obtain a method for estimating ϵ, which we illustrate by estimating ϵ_1. We consider the ratios from the last step for item 1 and all remaining items. If there are overall I items

$$\prod_{i \neq 1} \frac{p_{1i}}{p_{i1}} \approx \prod_{i \neq 1} \frac{\epsilon_1}{\epsilon_i} = \frac{\epsilon_1^{I-1}}{\prod_{i \neq 1} \epsilon_i}.$$

By including item 1 in both the numerator and denominator and using $\prod_i \epsilon_i = 1$, we get

$$= \frac{\epsilon_1^I}{\prod_i \epsilon_i} = \epsilon_1^I.$$

Therefore, we finally get

$$\hat{\epsilon}_1 = \left(\prod_{i \neq 1} \frac{p_{1i}}{p_{i1}} \right)^{1/I}.$$

C.4 Answers for Chapter 4

1. In this example, items 4 and 12 exhibit DIF, because their respective confidence ellipses do not cross the reference line.

2. At first sight it looks like all items exhibit DIF in this example, because none of the confidence ellipses crosses the reference line. This is a consequence of the underlying restriction.

   As explained in Section 4.2.4, the function for creating the graphical test in the `eRm` package, which was used for creating this plot, employs the restriction that the sum of all item parameters is zero in both groups. If DIF is balanced, this restriction (that corresponds to an anchor containing all items) works well. In this example, however, unbalanced DIF was simulated. Items 1 and 2 show strong DIF in the same direction. As a consequence, an approach that assumes balanced DIF will shift all other items such that the sum of all item parameters is zero again in both groups. This moves the other items away from the reference line.

   A more sensible way of interpreting a graphical test in this situation is to use an alternative reference line. Wright and Stone (1999), for example, suggest placing a line through the "major item stream". In this example, when we manually add a line through the majority of items (Figure C.1, dashed line), we can interpret more clearly that only items 1 and 2 show DIF. Note though that this procedure also makes an assumption, namely that it is the majority – rather than the minority – of items, that are DIF free. This is a very common assumption, but it can also be discussed (cf., e.g., Strobl et al., 2021).

3. Besides the graphical test we can use the item-wise Wald test for testing individual items for DIF. We need to select at least one

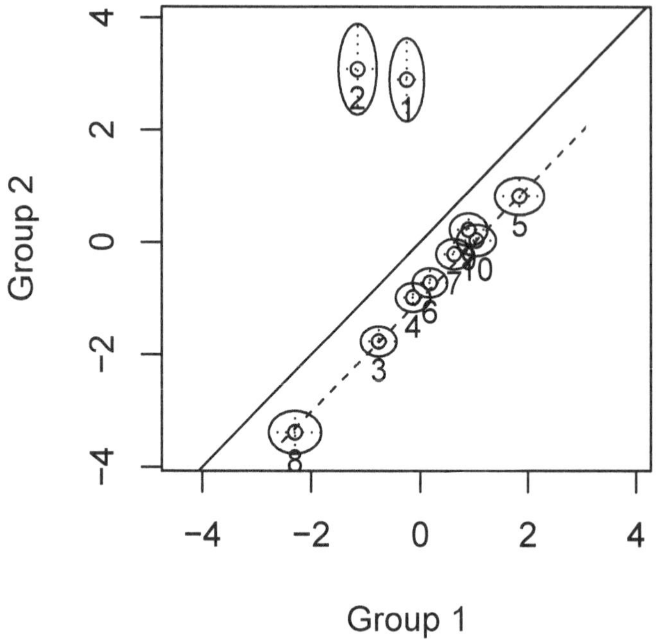

FIGURE C.1: Figure used in the solutions to the end-of-chapter exercises.

anchor item to place the item parameter estimates for both groups of test takers on the same scale so that we can compare them, but approaches using more than one anchor item have been shown to have better statistical power. Even if a method for testing individual items for DIF does not explicitly require the user to select anchor items, implicitly it will use some kind of anchoring (see also Exercise 2 and Section 6.2.5). This includes approaches that use the person parameter distributions for anchoring.

4. Suppose persons p and q fall into distinct groups between which DIF exists in item i but not item j. Let $\beta_i^{(p)}$ be the difficulty of the DIF item i for person p and all test takers in the same group as p. Let $\beta_i^{(q)}$ be the difficulty of the DIF item i for person q and all test takers in the same group as q. As a result of DIF, $\beta_i^{(p)} \neq \beta_i^{(q)}$ for the DIF item, item i.

   Using the argumentation from Section 2.4.3, for the DIF item i we obtain

$$\begin{aligned}\log(O_{pi}) - \log(O_{qi}) &= (\theta_p - \beta_i^{(p)}) - (\theta_q - \beta_i^{(q)}) \\ &= (\theta_p - \theta_q) + (\beta_i^{(q)} - \beta_i^{(p)})\end{aligned}$$

while for the non-DIF item j

$$\log(O_{pj}) - \log(O_{qj}) = (\theta_p - \beta_j) - (\theta_q - \beta_j) = \theta_p - \theta_q.$$

Since these two expressions are not equal, the multiplication condition does not hold.

5. A χ^2 goodness-of-fit test checks whether the Rasch model is able to predict, or account for, the observed data. Its reasoning is to compare the observed number of test takers showing each response pattern to the number that we would expect under the Rasch model. The differences between the observed and expected values are squared, and as a consequence, positive and negative differences in the sum will not cancel out. Then, each difference is scaled by the expected frequency. The resulting test statistic is compared against a χ^2 distribution, with large values of the statistic (or small p-values) indicating poor model fit. The problem with this approach is that in practice the expected frequencies of individual response pattern are too low to warrant a χ^2 distribution.
6. Rasch residuals are the differences between the observed responses and the expected values under the Rasch model. They are used for the calculation of the t and MSQ values of the infit and outfit statistics.
7. The key characteristic of the Q_3 statistic is that it is calculated for item pairs, rather than for individual items. It is also based on the calculation of model residuals, which are the differences between the observed responses and those expected under the Rasch (or any other IRT) model. The Q_3 statistic for a specific item pair is the correlation between the residuals of all test takers for one item with those for another item.
8. The test statistics of Ponocny (2001) have been suggested specifically for small samples. They are based on generating new response data that are equally plausible under the Rasch model as the observed data. The newly generated and the observed response data are compared by means of a test statistic, which allows an assessment of whether the data we have observed are plausible under the Rasch model or whether they indicate a model violation.
9. The solid ICC corresponds to the 2PL, and the dashed ICC corresponds to the 3PL. In the 3PL, the lower asymptote can be above zero to account for guessing (i.e., a probability of a correct answer above zero even for test takers with infinitely low abilities).
10. The slope of the item with the dashed ICC is larger than the slope of the item with the solid ICC. Thus, the item with the dashed ICC has the larger discrimination.

11. Models can be compared by means of likelihood ratio tests and information criteria, such as AIC and BIC. Likelihood ratio tests are only valid for nested models and restrictions that are not at the boundaries of the parameter space. Information criteria are more generally applicable, but provide only a descriptive comparison, no clear-cut decision like a significance test. Different information criteria can disagree about which model is to be preferred, because they use different amounts of penalization for the model complexity.

C.5 Answers for Chapter 5

1. To install and load the `psychotree` package

```
> install.packages("psychotree")
> library("psychotree")
```

2. To load the SPISA data

```
> data(SPISA, package = "psychotree")
```

3. We can get information about the structure by typing

```
> str(SPISA)
'data.frame': 1075 obs. of  6 variables:
$ spisa : num [1:1075, 1:45] 1 0 0 0 1 1 0 1 1 0 ...
..- attr(*, "dimnames")=List of 2
.. ..$ : NULL
.. ..$ : NULL
$ gender : Factor w/ 2 levels "female","male": 1 2 1 2
   2 2 2 2 2 1 ...
$ age : num 21 20 25 27 24 20 22 22 21 20 ...
$ semester: Ord.factor w/ 11 levels
   "1"<"2"<"3"<"4"<..: 3 1 9 10 8 1 2 2 1 1 ...
$ elite : Factor w/ 2 levels "no","yes": 1 1 1 1 1 2 1
   1 1 1 ...
$ spon : Ord.factor w/ 7 levels "never"<"<1/month"<..:
   3 6 3 1 4 3 1 4 4 3 ...
```

This shows that the SPISA data is a data frame containing in total 1075 observations and 6 variables. What is specific about the format of the SPISA data is that the item responses are stored like a single variable. This variable, named **spisa**, is actually a numeric matrix (we can see from the numbers in the squared brackets that is has rows 1 through 1075 and columns 1 through 45). The remaining five variables are covariates like gender (a factor), age (a numeric variable), semester (an ordered factor), etc.

4. The number of test takers and items are the numbers of rows and columns of **spisa**. We can display the dimensions of **spisa** by using

```
> dim(SPISA$spisa)
[1] 1075   45
```

   We see that **spisa** has 1075 rows and 45 columns. Thus, we have 1075 test takers and 45 test items.

5. To get the twelfth element from the eighteenth row, i.e., the response of test taker 18 to item 12, type

```
> SPISA$spisa[18, 12]
[1] 1
```

6. To get rows 65–70 of the tenth column, i.e., the responses of test takers 65 through 70 to item 10, type

```
> SPISA$spisa[65:70, 10]
```

7. To get all of test taker 422's responses

```
> SPISA$spisa[422, ]
 [1] 0 1 0 1 1 1 1 1 1 1 1 1 1 0 1 1 0 1 1 1 1 1 0 1 1
[26] 1 1 1 1 1 1 1 1 1 1 1 1 1 1 1 1 1 1 1 1
```

   There are 45 responses for each test taker, because the test has 45 items, as we have already noted above.

   If you want to verify this, one possible way to find out how many entries a vector contains is

```
> length(SPISA$spisa[422, ])
[1] 45
```

8. To select the responses for items 37 through 45 and store them

```
> natSci_responses <- SPISA$spisa[,37:45]
```

9. Either of the following will open the help file.

```
> ?rowSums
> help("rowSums")
```

The help file indicates that the **rowSums()** function computes the sum of each row of a matrix or data frame.

10. We can compute the test takers' sum scores with

```
> scores <- rowSums(natSci_responses)
```

11. To find out the highest number of items that test takers answered correctly, we can use different commands. For example, the summary command, that we have introduced above for summarizing the variables in a data set, is a generic command and can also be used to summarize the properties of the vector of sum scores we have created.

```
> summary(scores)
   Min. 1st Qu.  Median    Mean 3rd Qu.    Max.
  1.000   6.000   7.000   6.556   8.000   9.000
```

We find that the maximum number of items that test takers answered correctly is 9. This shows that there were individuals who answered all nine natural science items correctly. We also see that the minimum for the sum scores is 1. This means that there was no test taker who got all natural science items wrong, which would have resulted in a sum score of zero.

Other options for answering this question (that were not presented in Chapter 5) would be to use the command

```
> max(scores)
[1] 9
```

or to display the frequencies of the different sum scores in the form of a table or barplot.

```
> table(scores)
scores
  1   2   3   4   5   6   7   8   9
  6  10  30  68 135 228 272 221 105
> barplot(table(scores))
```

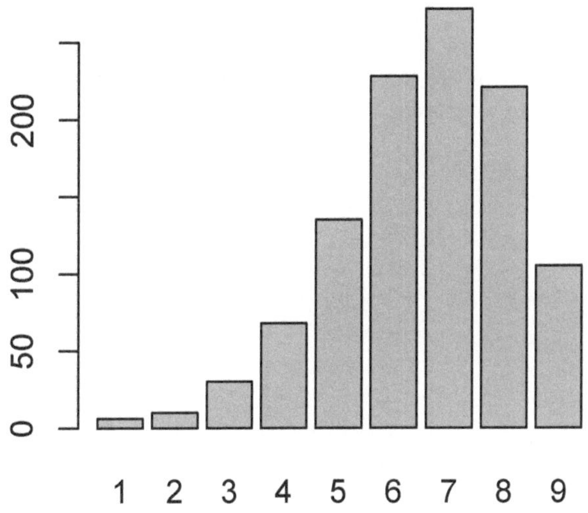

C.6 Answers for Chapter 6

1. To load the SPISA data, select the natural science items 37 through 45 and store them, type

```
> data(SPISA, package = "psychotree")
> natSci_responses <- SPISA$spisa[,37:45]
```

(In case you have not done so already for the previous exercises, you might need to install the **psychotree** package beforehand.)

To store the gender information in a separate vector, use

```
> gender <- SPISA$gender
```

Note that – in contrast to the FIMS data example from Chapter 6, where we have only included certain persons (i.e., certain rows, indicated by the first index in squared brackets) – here we are using the data from all persons. We have only restricted the columns (indicated by the second index in squared brackets) of the item response matrix to the natural science items. Therefore, we store the gender information for all test takers.

Remember that it would also be possible to proceed with the vector for gender being part of the **SPISA** data frame. Then we would have to refer to it by **SPISA$gender** in the following exercises.

2. We can load the package, estimate the Rasch model, and summarize the results using the code below. From the ICCs and the summary tables, we can see that item 9 is the easiest item, with a difficulty of −1.28, and that item 3 is the hardest item, with a difficulty of 1.52.

```
> library("eRm")
> rm_SPISA <- RM(natSci_responses)
> summary(rm_SPISA)

Results of RM estimation:

Call:  RM(X = natSci_responses)

Conditional log-likelihood: -3100.61
Number of iterations: 15
Number of parameters: 8

Item (Category) Difficulty Parameters (eta): with 0.95 CI:
   Estimate Std. Error lower CI upper CI
I2    0.114      0.069   -0.020    0.249
I3    1.520      0.065    1.393    1.646
I4   -0.565      0.080   -0.722   -0.409
I5    0.973      0.063    0.849    1.097
I6   -0.266      0.074   -0.411   -0.121
I7   -0.118      0.072   -0.259    0.022
I8    0.332      0.066    0.202    0.462
I9   -1.280      0.100   -1.475   -1.085

Item Easiness Parameters (beta) with 0.95 CI:
        Estimate Std. Error lower CI upper CI
beta I1    0.709      0.083    0.546    0.872
beta I2   -0.114      0.069   -0.249    0.020
beta I3   -1.520      0.065   -1.646   -1.393
beta I4    0.565      0.080    0.409    0.722
beta I5   -0.973      0.063   -1.097   -0.849
beta I6    0.266      0.074    0.121    0.411
beta I7    0.118      0.072   -0.022    0.259
beta I8   -0.332      0.066   -0.462   -0.202
beta I9    1.280      0.100    1.085    1.475
```

Which items are easy and which are hard is also visible from a plot of all ICCs:

```
> plotjointICC(rm_SPISA)
```

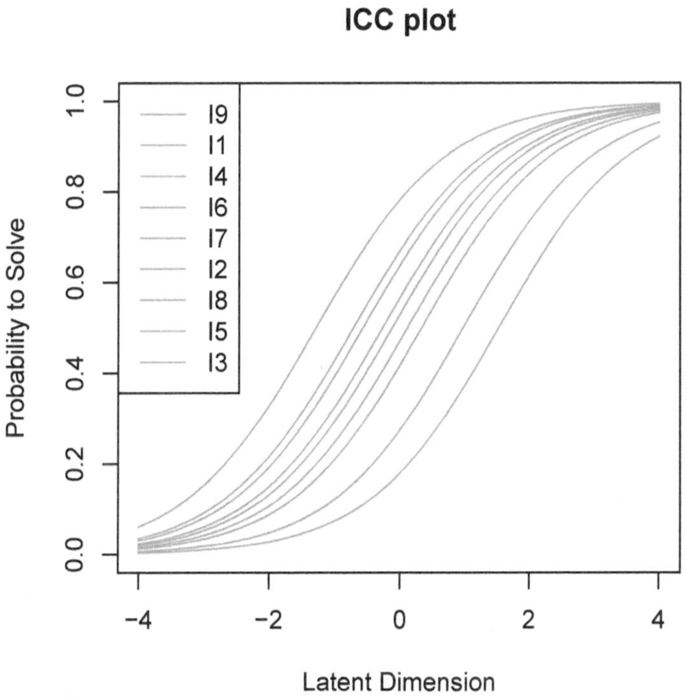

3. To create the person item map, type

   ```
   > plotPImap(rm_SPISA)
   ```

   (The plot is displayed on the following page.)

   The abilities and difficulties should ideally be centered in the same area and show a large overlap. Here, we find that the person distribution is somewhat skewed, with more test takers showing high abilities. These smart test takers provide a lot of information for estimating the item parameters especially for the more difficult items. This leads to smaller standard errors and confidence ellipses for the difficult items compared to the easy items (cf. output for previous and next exercise).

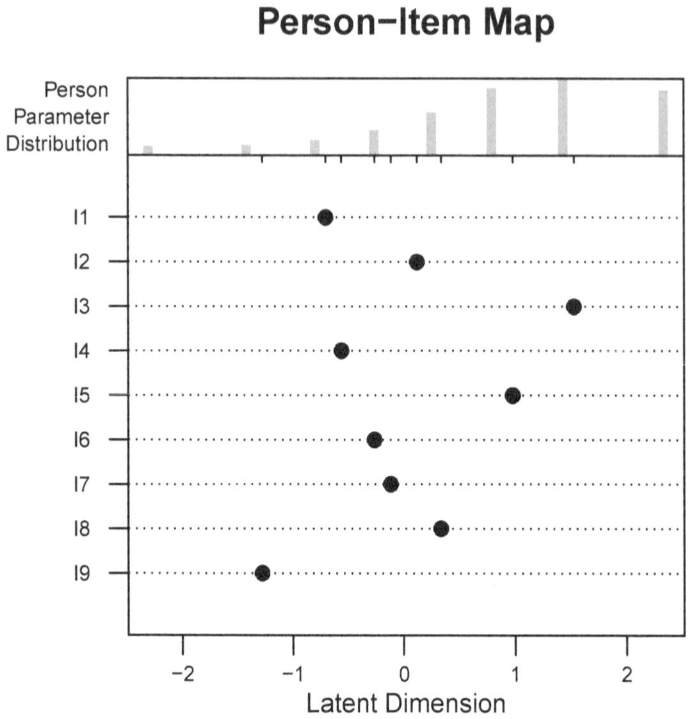

4. The code below will perform the likelihood ratio test for all items jointly. It indicates that there is a significant difference in item parameter estimates between female and male test takers (which were the only genders recorded for this data set).

```
> lrt_SPISA <- LRtest(rm_SPISA, splitcr = gender)
> lrt_SPISA

Andersen LR-test:
LR-value: 119.083
Chi-square df: 8
p-value:  0
```

5. The code below will produce the plot for the graphical test.

```
> plotGOF(lrt_SPISA,
+         tlab = "item",
+         conf = list(gamma = 0.95, col = 1))
```

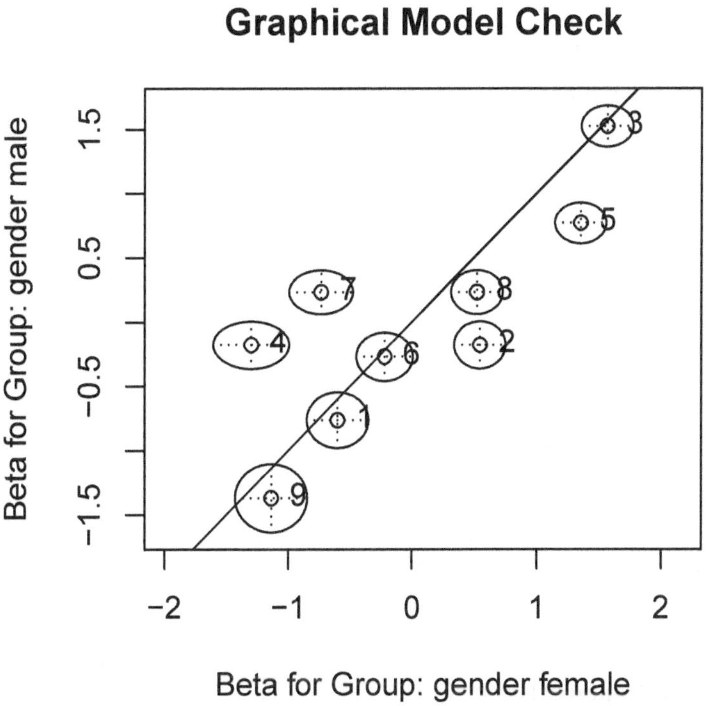

The plot suggests that items 4 and 7 are harder to solve for males, items 2 and 5 (and borderline item 8) are harder for females.

If you are interested, you can look up the item content on the `psychotree` help page for the `SPISA` data.

6. The following code will run an item-wise Wald test.

```
> Waldtest(rm_SPISA, splitcr = gender)

Wald test on item level (z-values):

        z-statistic p-value
beta I1      -0.962   0.336
beta I2      -5.145   0.000
beta I3      -0.319   0.750
beta I4       6.131   0.000
beta I5      -4.387   0.000
beta I6      -0.334   0.738
beta I7       6.117   0.000
beta I8      -2.133   0.033
beta I9      -1.155   0.248
```

The result confirms that items 2, 4, 5, 7, and 8 show DIF with respect to gender.

7. To eliminate one item at a time use

```
> stepwiseIt(rm_SPISA, criterion = list("Waldtest", gender))
Eliminated item - Step 1: I4
Eliminated item - Step 2: I7
Eliminated item - Step 3: I2
Eliminated item - Step 4: I5

Results for stepwise item elimination:
Number of steps: 4
Criterion: Waldtest

           z-statistic p-value
Step 1: I4       6.131   0.000
Step 2: I7       7.061   0.000
Step 3: I2      -3.056   0.002
Step 4: I5      -2.574   0.010
```

To load the **psychotools** package and conduct a Wald test for gender with the iterative forward mean test statistic threshold anchoring approach, use

```
> library("psychotools")
> anchortest(natSci_responses ~ gender,
+            class = "forward",
+            select = "MTT")$final_tests

 Simultaneous Tests for General Linear Hypotheses

Linear Hypotheses:
                      Estimate Std. Error z value
natSci_responses1 == 0  0.02549    0.15003   0.170
natSci_responses2 == 0  0.59011    0.16624   3.550
natSci_responses4 == 0 -1.25871    0.21258  -5.921
natSci_responses5 == 0  0.44585    0.15624   2.854
natSci_responses6 == 0 -0.08611    0.13735  -0.627
natSci_responses7 == 0 -1.10438    0.18513  -5.965
natSci_responses8 == 0  0.15455    0.12643   1.222
natSci_responses9 == 0  0.09591    0.23311   0.411
                           Pr(>|z|)
natSci_responses1 == 0     0.865069
natSci_responses2 == 0     0.000385 ***
natSci_responses4 == 0 0.00000000320 ***
natSci_responses5 == 0     0.004321 **
```

```
natSci_responses6 == 0       0.530712
natSci_responses7 == 0 0.0000000244 ***
natSci_responses8 == 0       0.221558
natSci_responses9 == 0       0.680765
---
Signif. codes:
0 '***' 0.001 '**' 0.01 '*' 0.05 '.' 0.1 ' ' 1
(Univariate p values reported)
```

In this example, DIF seems to be almost balanced so that the conclusions after purification and anchoring are very similar to those of the naive DIF tests using the sum zero constraint: Items 2, 4, 5, and 7 show significant DIF with respect to gender in each approach. The results differ only with respect to item 8, that showed (borderline) DIF in the naive tests, but is not flagged by the purification and anchoring approaches.

8. We can calculate the infit and outfit statistics using the following code.

```
> itemfit(person.parameter(rm_SPISA))

Itemfit Statistics:
      Chisq  df p-value Outfit MSQ Infit MSQ Outfit t
I1  840.580 969   0.999      0.867     0.932   -1.621
I2  928.806 969   0.819      0.958     0.956   -0.839
I3 1084.479 969   0.006      1.118     1.094    2.855
I4  827.622 969   1.000      0.853     0.941   -1.965
I5  888.542 969   0.969      0.916     0.934   -2.520
I6  859.388 969   0.995      0.886     0.941   -1.822
I7  993.474 969   0.286      1.024     1.007    0.430
I8  938.468 969   0.754      0.967     0.976   -0.734
I9  692.359 969   1.000      0.714     0.876   -2.660
   Infit t Discrim
I1  -1.266   0.224
I2  -1.278   0.249
I3   3.507  -0.154
I4  -1.186   0.251
I5  -2.677   0.275
I6  -1.419   0.260
I7   0.188   0.122
I8  -0.763   0.217
I9  -1.723   0.311
```

We find t values higher above 2 for item 3 and below -2 for items 5 and 9.

For these items we also plot the empirical ICCs:

```
> plotICC(rm_SPISA, item.subset=c(3,5,9), empICC =
+       list("raw"), empCI = list())
```

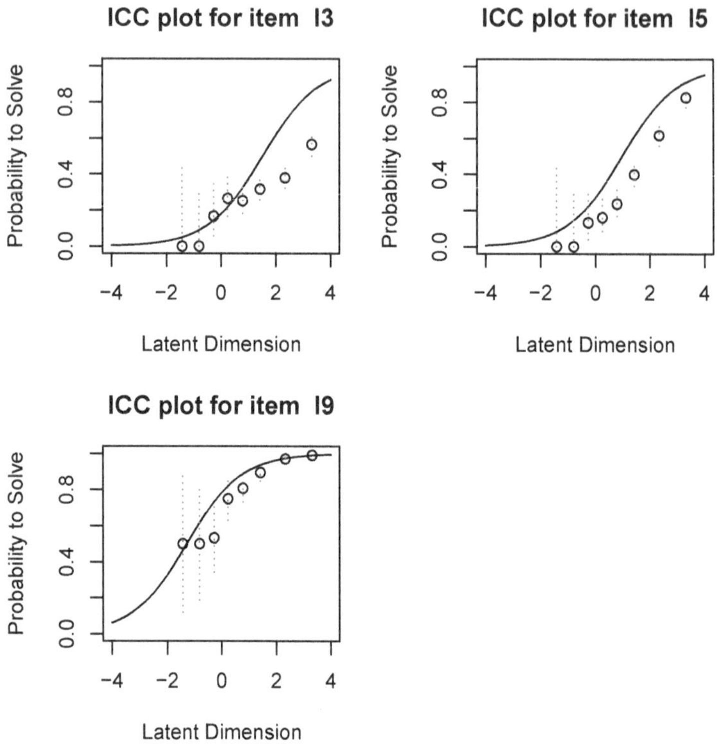

According to the fit statistics and in line with the shape of the empirical ICC, item 3 shows underfit. Its empirical ICC is flatter than expected. Items 5 and 9 show overfit according to the fit statistics. The empirical ICC of item 9 shows a steeper slope between the third and fourth point of the empirical ICC, which is in line with overfit. It also seems to flatten out to the left, which could indicate guessing. For item 5 the empirical ICC is inconclusive. It does show a small jump between the second and third point of the empirical ICC, which could indicate overfit. However, note that for all items the points for the low abilities on the left represent responses from very few test takers, as indicated by the wide confidence intervals.

9. To estimate and print the person parameters for all persons use the following commands.

```
> theta <- person.parameter(rm_SPISA)
> summary(theta)
Estimation of Ability Parameters

Collapsed log-likelihood: -28.60929
Number of iterations: 9
Number of parameters: 8

ML estimated ability parameters (without spline
 interpolated values):
               Estimate Std. Err.      2.5 %      97.5 %
theta P1      1.4223019 0.8513794 -0.2463712  3.0909749
theta P2      2.3343600 1.1017470  0.1749755  4.4937446
theta P3      1.4223019 0.8513794 -0.2463712  3.0909749
theta P4      1.4223019 0.8513794 -0.2463712  3.0909749
theta P5      1.4223019 0.8513794 -0.2463712  3.0909749
theta P6      1.4223019 0.8513794 -0.2463712  3.0909749
theta P7      0.2422377 0.7205098 -1.1699356  1.6544109
theta P8      0.7842306 0.7583361 -0.7020808  2.2705421
theta P9      1.4223019 0.8513794 -0.2463712  3.0909749
theta P10     0.2422377 0.7205098 -1.1699356  1.6544109
theta P12    -0.8037493 0.7493523 -2.2724528  0.6649542
theta P13    -2.3100310 1.0881571 -4.4427797 -0.1772823
theta P14     0.7842306 0.7583361 -0.7020808  2.2705421
theta P17     1.4223019 0.8513794 -0.2463712  3.0909749
theta P18     0.7842306 0.7583361 -0.7020808  2.2705421
theta P19    -0.8037493 0.7493523 -2.2724528  0.6649542
```

The output is again truncated to save space. Several persons, for example persons 1, 3, 4, 5, and 6 receive exactly the same ability estimate and standard error. This is due to the fact that these persons have worked on the same items and scored the same sum score.

We also see that some persons, for example persons 11, 15, and 16, do not receive an estimate in this output. This is the case for persons who solved all items or none of the items correctly.

In the case of persons 11, 15, and 16, we can look up the respective sum scores

```
> rowSums(natSci_responses)[c(11, 15, 16)]
[1] 9 9 9
```

and find that these test takers have indeed solved all nine natural science items correctly. When we print out the person parameter estimates for each sum score via

```
> theta

Person Parameters:

 Raw Score    Estimate Std.Error
         1 -2.3100310 1.0881571
         2 -1.4239417 0.8382345
         3 -0.8037493 0.7493523
         4 -0.2707924 0.7173574
         5  0.2422377 0.7205098
         6  0.7842306 0.7583361
         7  1.4223019 0.8513794
         8  2.3343600 1.1017470
         9  3.3135660        NA
```

we can also see that for test takers who answered all nine items correctly, the ability estimate has no standard error. The reason is that it does not result from maximum likelihood estimation like the other estimates, but from spline interpolation.

C.7 Answers for Chapter 7

1. To load the SPISA data, select the natural science items 37 through 45, store them, and assign the item names, type

```
> data(SPISA, package = "psychotree")
> natSci_responses <- SPISA$spisa[,37:45]
> colnames(natSci_responses) <- paste("I", 1:9)
```

2. After loading the package, in principle, we can fit the three models as follows.

```
> library("mirt")
> mirt_rm_SPISA <- mirt(natSci_responses, 1, "Rasch")
> mirt_2pl_SPISA <- mirt(natSci_responses, 1, "2PL")
> mirt_3pl_SPISA <- mirt(natSci_responses, 1, "3PL")
```

However, you will notice that for this data set the default number of 500 iterations is not enough for the 3PL model to converge. In this case, `mirt` will issue a warning like: "EM cycles terminated after 500 iterations".

Therefore, we increase the number of iterations (or cycles) to 5000:

```
> mirt_3pl_SPISA <- mirt(natSci_responses, 1, "3PL",
+                        technical = list(NCYCLES = 5000))
```

3. We compare the Rasch and 2PL models using

```
> anova(mirt_rm_SPISA, mirt_2pl_SPISA)

Model 1: mirt(data = natSci_responses, model = 1,
 itemtype = "Rasch")
Model 2: mirt(data = natSci_responses, model = 1,
 itemtype = "2PL")

       AIC    SABIC       HQ      BIC    logLik     X2
1 10175.78 10193.82 10194.64 10225.58 -5077.891    NaN
2 10161.68 10194.16 10195.64 10251.33 -5062.843 30.098
   df   p
1 NaN NaN
2   8   0
```

For this first model comparison, the significant result of the likelihood ratio test indicates that the 2PL provides a better account of the data than the Rasch model. This agrees with the AIC, although the BIC (which penalizes the number of parameters more strongly) prefers the Rasch model.

We can then compare the 2PL and 3PL models using

```
> anova(mirt_2pl_SPISA, mirt_3pl_SPISA)

Model 1: mirt(data = natSci_responses, model = 1,
 itemtype = "2PL")
Model 2: mirt(data = natSci_responses, model = 1,
 itemtype = "3PL", technical = list(NCYCLES = 5000))

       AIC    SABIC       HQ      BIC    logLik    X2
1 10161.68 10194.16 10195.64 10251.33 -5062.843   NaN
2 10175.37 10224.07 10226.29 10309.83 -5060.683 4.319
   df     p
1 NaN   NaN
2   9 0.889
```

For the second comparison, remember that the likelihood ratio test should not be used for comparing 2PL and 3PL models because the tested restriction is at the boundary of the parameter space. Instead, we concentrate on the information criteria, which all prefer the 2PL model. Overall, this indicates that – from the three models we have investigated – the 2PL model provides the best account of the data. This may come as a surprise because the empirical ICC for the exemplary item 9 in the exercises to the previous chapter does look like guessing was involved. However, information criteria do not rely on fit alone, but on balancing fit against complexity – and according to the results, the 2PL model provides the best compromise between the two for these data.

Since the natural science items consist of a combination of multiple choice and open questions, it would also be possible to use different models for the two item types (cf. Section 4.5.1.2). In `mirt` this is possible by means of specifying as `itemtype` a vector with one entry for each item instead of a single model for all items (this was not asked in the exercise). However, in our example using the 2PL model for all items works best.

4. We can plot the ICCs as follows:

```
> plot(mirt_2pl_SPISA, type = "trace")
```

(The plot is displayed on the following page.)

In the exercises for the previous chapter, the in/outfit statistics and the empirical ICCs indicated that item 3 showed underfit and items 5 and 9 showed overfit when all items were restricted to the same slope in the Rasch model.

The expected ICCs for the 2PL model are very plausible for item 3, which shows a much lower slope than the other items now. When we look carefully, we also see that items 5 and 9 are among those with a higher slope. Overall, the shapes of the 2PL ICCs are in line with the previous results.

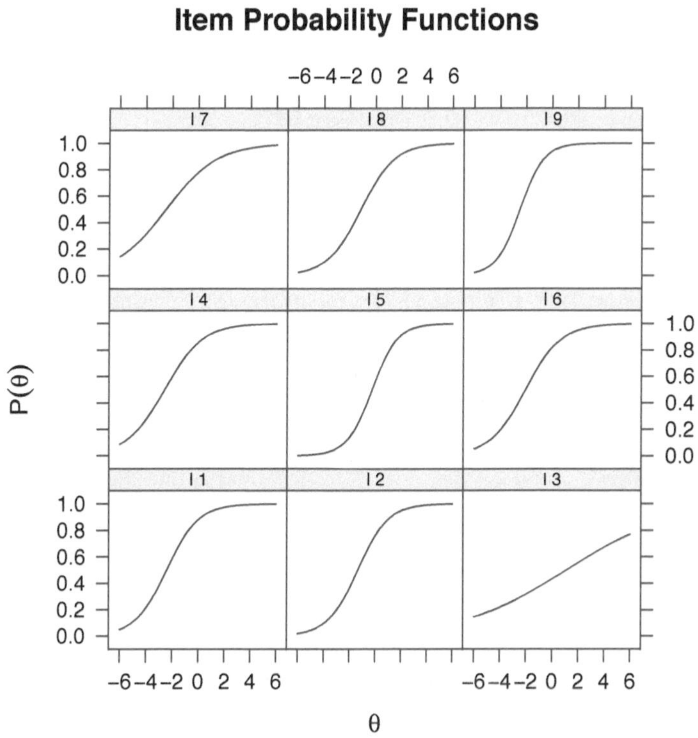

5. In order to display the confidence intervals, we need to fit the model with the option SE = TRUE. In case you have not already used this option in the previous exercises, we need to refit the model. Then we can show the item parameter estimates together with their confidence intervals.

```
> mirt_2pl_SPISA <- mirt(natSci_responses, 1,
+                        "2PL", SE = TRUE)
> coef(mirt_2pl_SPISA, IRTpars = TRUE)


$`I 1`
            a      b  g  u
par     0.825 -2.392  0  1
CI_2.5  0.509 -3.152 NA NA
CI_97.5 1.140 -1.632 NA NA

$`I 2`
            a      b  g  u
par     0.856 -1.332  0  1
CI_2.5  0.572 -1.715 NA NA
```

```
CI_97.5 1.141 -0.948 NA NA

$`I 3`
            a     b  g  u
par     0.248 1.155  0  1
CI_2.5  0.058 0.163 NA NA
CI_97.5 0.439 2.148 NA NA

$`I 4`
            a      b  g  u
par     0.684 -2.580  0  1
CI_2.5  0.396 -3.529 NA NA
CI_97.5 0.972 -1.631 NA NA

$`I 5`
            a      b  g  u
par     1.033 -0.250  0  1
CI_2.5  0.708 -0.402 NA NA
CI_97.5 1.358 -0.097 NA NA

$`I 6`
            a      b  g  u
par     0.731 -2.030  0  1
CI_2.5  0.455 -2.697 NA NA
CI_97.5 1.008 -1.364 NA NA

$`I 7`
            a      b  g  u
par     0.511 -2.489  0  1
CI_2.5  0.273 -3.571 NA NA
CI_97.5 0.749 -1.407 NA NA

$`I 8`
            a      b  g  u
par     0.773 -1.157  0  1
CI_2.5  0.509 -1.523 NA NA
CI_97.5 1.036 -0.791 NA NA

$`I 9`
            a      b  g  u
par     1.097 -2.493  0  1
CI_2.5  0.680 -3.196 NA NA
CI_97.5 1.513 -1.790 NA NA
```

```
$GroupPars
        MEAN_1 COV_11
par          0      1
CI_2.5      NA     NA
CI_97.5     NA     NA
```

The item with the lowest discrimination parameter (in column `a`) is item 3. We already saw this very clearly in the ICC plot in the previous exercise.

The two items with the highest slopes are items 5 and 9. These are exactly the items that were diagnosed with overfit in the exercises for the previous chapter, where all items were restricted to the same slope by the Rasch model. This was less obvious from the ICC plot.

The confidence intervals for the item parameters are all inconspicuous. In particular, all of the confidence intervals for the slope parameters are clearly in the positive range.

6. M_2, $S\text{-}X_i^2$, and the fit statistic of Stone (2000) can be computed with the following commands:

```
> M2(mirt_2pl_SPISA)
            M2 df           p      RMSEA    RMSEA_5
stats 47.71036 27 0.008270283 0.02672451 0.01348389
        RMSEA_95      SRMSR       TLI       CFI
stats 0.03892722 0.03387252 0.9017438 0.9263079
```

```
> itemfit(mirt_2pl_SPISA)
  item   S_X2 df.S_X2 RMSEA.S_X2 p.S_X2
1  I 1  9.641       5      0.029  0.086
2  I 2  9.153       5      0.028  0.103
3  I 3  4.068       5      0.000  0.540
4  I 4 13.289       6      0.034  0.039
5  I 5  0.844       4      0.000  0.932
6  I 6 10.654       5      0.032  0.059
7  I 7  3.071       5      0.000  0.689
8  I 8  2.993       5      0.000  0.701
9  I 9  7.398       6      0.015  0.286
```

```
> set.seed(2345)
> itemfit(mirt_2pl_SPISA, fit_stats = "X2*")
  item X2_star p.X2_star
1  I 1   0.209     0.815
2  I 2   0.268     0.337
```

```
3  I 3  0.126      0.510
4  I 4  0.158      0.776
5  I 5  0.119      0.799
6  I 6  0.446      0.124
7  I 7  0.268      0.281
8  I 8  0.203      0.492
9  I 9  0.490      0.501
```

We find that the p-value for the global fit statistic M_2 is below 0.05, indicating a poor fit. On the other hand, the upper bound of the confidence interval for the RMSEA as well as the SRMSR fall below 0.05, suggesting that the model does provide a good account of the data. This is an example where different fit statistics produce contradictory results. One possible reason could be that the misfit is not substantial, but based on the large sample size the p-value for the M_2 statistic still turns out significant.

The approximate p-values for the itemwise S-X_i^2 statistic indicate that all items except for item 4 show an acceptable fit for the 2PL. The fit statistic of Stone (2000) does not flag any item for misfit. Again, different fit statistics may disagree, and items flagged by certain statistics should be investigated by content experts, but should not be excluded based solely on fit statistics.

7. To estimate the person parameters by means of the maximum a posteriori (MAP) approach and display the first few entries

```
> theta_map <- fscores(mirt_2pl_SPISA, method = "MAP",
+                       full.scores.SE = TRUE)
> head(theta_map)
              F1     SE_F1
[1,]  0.1398640 0.7423583
[2,]  0.7430083 0.7878267
[3,]  0.1398640 0.7423583
[4,]  0.1102551 0.7403563
[5,] -0.1202192 0.7257094
[6,]  0.1102551 0.7403563
```

We notice that the first and the third person, as well as the fourth and the sixth person, receive exactly the same point estimate and standard error. While in the Rasch model all persons with the same sum score receive the same estimate, in the 2PL model it is persons with the same response pattern who receive the same estimate. We can verify this by looking into the item responses (this was not asked in the exercise).

```
> cbind(head(natSci_responses), head(rowSums(
+       natSci_responses)))
     I 1 I 2 I 3 I 4 I 5 I 6 I 7 I 8 I 9
[1,]   1   1   0   1   0   1   1   1   1 7
[2,]   1   1   0   1   1   1   1   1   1 8
[3,]   1   1   0   1   0   1   1   1   1 7
[4,]   0   1   1   1   1   1   0   1   1 7
[5,]   1   1   1   1   0   0   1   1   1 7
[6,]   0   1   1   1   1   1   0   1   1 7
> rbind(natSci_responses[1, ],natSci_responses[3, ])
     I 1 I 2 I 3 I 4 I 5 I 6 I 7 I 8 I 9
[1,]   1   1   0   1   0   1   1   1   1
[2,]   1   1   0   1   0   1   1   1   1
> rbind(natSci_responses[4, ],natSci_responses[6, ])
     I 1 I 2 I 3 I 4 I 5 I 6 I 7 I 8 I 9
[1,]   0   1   1   1   1   1   0   1   1
[2,]   0   1   1   1   1   1   0   1   1
```

While several persons have scored a sum score of 7, only those two pairs of persons who have scored exactly the same response pattern receive the same ability estimate.

C.8 Answers for Chapter 8

1. To load the SPISA data, select the natural science items 37 through 45, store them, and assign the item names, type

```
> data(SPISA, package = "psychotree")
> natSci_responses <- SPISA$spisa[,37:45]
> colnames(natSci_responses) <- paste("I", 1:9)
```

2. After loading the package, we can fit the Rasch model and display the item parameters.

```
> library("TAM")
> tam_rm_SPISA <- tam.mml(natSci_responses)
> tam_rm_SPISA$item
```

```
    item    N         M   xsi.item AXsi_.Cat1 B.Cat1.Dim1
I 1  I 1 1075 0.8520930 -1.9164874 -1.9164874           1
I 2  I 2 1075 0.7293023 -1.0957379 -1.0957379           1
I 3  I 3 1075 0.4297674  0.3147499  0.3147499           1
I 4  I 4 1075 0.8344186 -1.7735493 -1.7735493           1
I 5  I 5 1075 0.5525581 -0.2343042 -0.2343042           1
I 6  I 6 1075 0.7925581 -1.4756826 -1.4756826           1
I 7  I 7 1075 0.7693023 -1.3282495 -1.3282495           1
I 8  I 8 1075 0.6883721 -0.8779046 -0.8779046           1
I 9  I 9 1075 0.9079070 -2.4835889 -2.4835889           1
```

We find again that item 3 has the highest and item 9 has the lowest difficulty (cf. exercises for Chapter 6).

3. Table 9.1 indicates that the following function plots the ICCs (a.k.a. item response functions or trace lines) for a fitted model.

```
> IRT.irfprobPlot(tam_rm_SPISA)
```

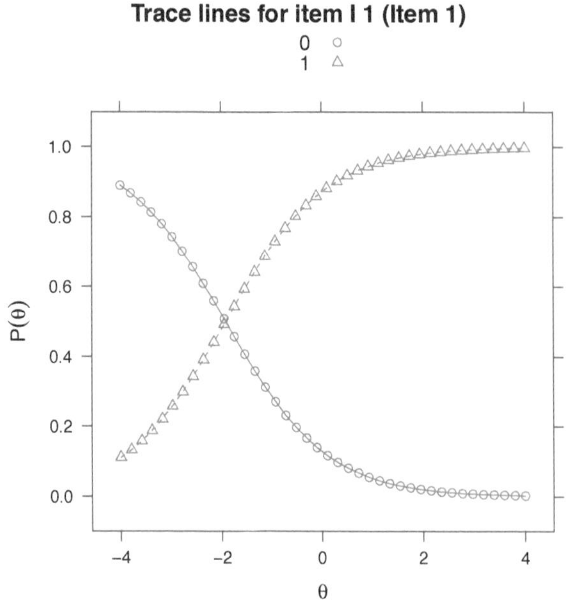

Here we display the resulting plot only for the first item. The IRT.irfprobPlot function in TAM plots the predicted probabilities of both responses, zero (for incorrect or negative responses) and one (for correct or positive responses). The curve for the correct responses (indicated by triangles and rising to the right) corresponds to the expected ICC in the respective plots in the previous chapters.

4. To compute and display the SRMSR

```
> rm_gof <- IRT.modelfit(tam_rm_SPISA)
**** Calculate Residuals
**** Calculate Counts
**** Calculate Covariances
> rm_gof$fitstat
100*MADCOV        SRMR       SRMSR
0.66637079 0.03799329 0.04682207
```

The SRMSR of approx. 0.0468 does not exceed the cutoff of 0.05 suggested by Maydeu-Olivares (2013). (Still, we found in the exercises to the previous chapter that a 2PL model provides an even better fit for this data set.)

5. The person parameters estimated by means of the expected a posteriori (EAP) approach can be extracted directly from the fitted model.

```
> head(tam_rm_SPISA$person)
  pid case pweight score max       EAP    SD.EAP
1   1    1       1     7   9 0.1133536 0.5410890
2   2    2       1     8   9 0.4136751 0.5553007
3   3    3       1     7   9 0.1133536 0.5410890
4   4    4       1     7   9 0.1133536 0.5410890
5   5    5       1     7   9 0.1133536 0.5410890
6   6    6       1     7   9 0.1133536 0.5410890
```

For this, as well as for other estimation approaches, we could also use the function `IRT.factor.scores()`, cf. Table 9.1.

We find again that with the Rasch model, persons 1, 3, 4, 5, and 6 receive exactly the same ability estimate and standard error. This is due to the fact that these persons have worked on the same items and scored the same sum score.

C.9 Answers for Chapter 9

1. To load the SPISA data, select the natural science items 37 though 45, store them, and assign the item names, type

```
> data(SPISA, package = "psychotree")
> natSci_responses <- SPISA$spisa[, 37:45]
> colnames(natSci_responses) <- paste("I", 1:9)
```

2. To load the **rstan** package, enter

```
> library("rstan")
```

   You will need to configure your R installation to be able to compile C++ code and change R's current working directory to the location where you saved `rasch.stan`. Recall you can view R's current working directory by entering `getwd()`, and you can set it using `setwd()` or RStudio's menu to select the path to the correct folder.

3. We need to give three pieces of data to Stan: the number of test takers, the number of items and the data matrix. The number of test takers is the number of rows of the SPISA data and the number of test items is the number of columns. Thus, we can fit the model by entering the following.

```
> SPISA_args <- list(
+   num_person = nrow(natSci_responses),
+   num_item = ncol(natSci_responses),
+   U = natSci_responses
+ )
```

```
> stan_rm_SPISA <- stan("rasch.stan", data = SPISA_args,
+                       iter = 4000)
```

   Here we have already set the number of iterations to 4000 to ensure that our effective sample size is large enough.

4. We can use the `check_rhats()` function from Section 9.2 to determine whether the sampler has converged. After copying that function into R

```
> check_rhats <- function(stan_fit) {
+   stats <- summary(stan_fit)
+   all(stats$summary[, "Rhat"] < 1.1)
+ }
```

   we can check whether the sampler converged by entering

```
> check_rhats(stan_rm_SPISA)
[1] TRUE
```

This function returns **TRUE**, meaning we have no reason to suspect that our samples are not representative of the posterior distribution.

5. We can print summary statistics for the item parameters as follows.

```
> print(stan_rm_SPISA, pars = "beta")
Inference for Stan model: rasch.
4 chains, each with iter=4000; warmup=2000; thin=1;
post-warmup draws per chain=2000, total post-warmup
 draws=8000.

         mean se_mean   sd  2.5%   25%   50%   75% 97.5% n_eff Rhat
beta[1] -1.91       0 0.09 -2.10 -1.97 -1.91 -1.85 -1.73  6957    1
beta[2] -1.10       0 0.08 -1.25 -1.15 -1.10 -1.05 -0.95  8281    1
beta[3]  0.31       0 0.07  0.18  0.26  0.31  0.35  0.44  9106    1
beta[4] -1.77       0 0.09 -1.95 -1.83 -1.77 -1.71 -1.60  6700    1
beta[5] -0.24       0 0.07 -0.38 -0.29 -0.24 -0.19 -0.11  7985    1
beta[6] -1.48       0 0.08 -1.64 -1.53 -1.48 -1.42 -1.31  7499    1
beta[7] -1.33       0 0.08 -1.48 -1.38 -1.33 -1.28 -1.17  7111    1
beta[8] -0.88       0 0.07 -1.02 -0.93 -0.88 -0.83 -0.74  7726    1
beta[9] -2.47       0 0.11 -2.68 -2.54 -2.47 -2.40 -2.26  6479    1
```

The easiest item will have the lowest mean, indicating that Item 9 is the least difficult. The hardest item will have the highest mean, indicating that Item 3 is the most difficult.

6. We can use the output from the last question to determine both HPD intervals. We use the 2.5% and 97.5% percentiles, cutting off 95% in between them, to determine that the 95% HPD interval for the difficulty of Item 3 runs from 0.18 to 0.44. We can use the 25% and 75% percentiles, cutting off 50% in between them, to determine that the 50% HPD interval runs from 0.26 to 0.35. You will sometimes see 50% HPD intervals instead of 95% HPD intervals with posterior predictive checks, as they are more diagnostic in some cases.

7. For the following steps first copy and run the definitions for the functions **biserial_cor()**, **sample_biserial_ppd()**, and **plot_post_pred_items()** in R (not shown). We can then plot the observed point biserial correlations against their corresponding posterior predictive distribution using the approach from Section 9.3, shown below.

```
> obs_cor <- biserial_cor(natSci_responses)
> pred_cor <- sample_biserial_ppd(stan_rm_SPISA)
> plot_post_pred_items(obs_cor, pred_cor)
```

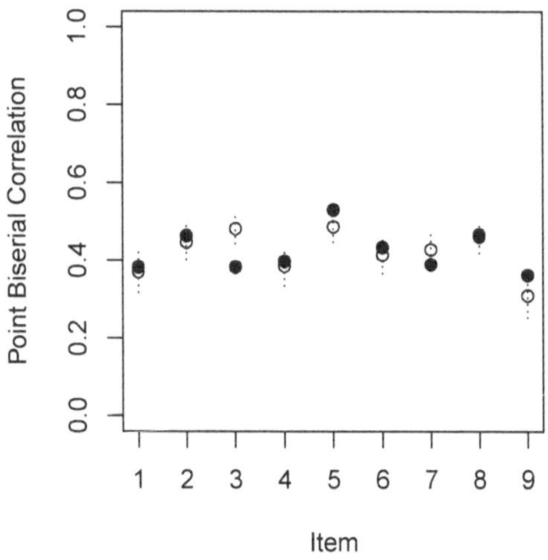

Most of the observed point biserial correlations fall within 95% HPD intervals for their values. Only the third item falls below its HPD interval.

95% HPD intervals should contain the observed values 95% of the time, so a single value falling outside of its corresponding interval does not provide strong evidence against the Rasch model per se. For this data set, however, we already saw in the exercises to the previous chapter (p. 277) that the third item has a notably lower slope than the other items when fitted with the 2PL model, which is in line with this result.

C.10 Answers for Chapter 10

1. The weights w_{ia} in the linear logistic test model determine which cognitive operations are required by an item. Typically, the weights for each item are specified a priori by content experts during the item construction phase.
2. The main difference that distinguishes mixture Rasch models from model-based recursive partitioning and explanatory Rasch models is that mixture models use latent classes, while model-based recursive partitioning and explanatory Rasch models use observed covariates to detect differences in the parameters of different groups of test takers. For latent classes both the number of classes and the class membership of the test takers are unknown, while, for example, for

categorical observed covariates both the number of categories and the value of this covariate for each test taker are captured in the data.

3. It is possible – but not always a good idea – to create a joint score from two or more separate dimensions. A joint sum score or average result is often reported, e.g., for attainment tests with several sub-dimensions or for school grades. What one should be aware of, however, is that whenever a joint score is reported, some form of weighting has been involved. This weighting could be such that all sub-dimensions count equally, or it could weight some higher than others, but either way a decision on the weighting has to be made.

   For more distinct abilities as well as personality scales, typically the different dimensions are not aggregated but reported separately to show the full performance or personality profile. A visual representation that tries to still allow capturing the results at one glance, e.g., is a radial display where the extent that the test takers scored on each trait corresponds to one spike of a star-like shape.

C.11 Answers for Chapter 11

1. The Partial Credit and the Rating Scale model, like the Rasch model, provide sufficient statistics for the person parameters. These act as a substitute for the person parameters, allowing us to estimate only the item parameters in conditional maximum likelihood estimation. (They can also be estimated by means of marginal maximum likelihood and Bayesian estimation.)
2. Of the four CCCs for categories 0, 1, 2, and 3, the two in the middle – for categories 1 and 2 – would be bell-shaped. The one for the lowest category 0 would be decreasing in ability. The one for the highest category 3 would be increasing in ability. The reason why the CCCs for the extreme categories have these monotone shapes is that they are the only options after a certain point on the ability scale. Remember that for each ability value the heights of the CCCs add up to one. In the middle of the continuum, we want the probability to score in a higher category to take over that of a lower category with increasing ability. For this, we need the bell-shaped CCCs for the middle categories. For the extreme categories, however, there is no successor category. For even higher (or lower) abilities, the probability of scoring in the highest (or lowest) category approaches 1, as the probabilities for all other categories approach 0.

3. The first professor, who graded all of the question herself, is the more likely candidate for successfully using the Rating Scale model. The four different teaching assistants of the second professor are much less likely to have each applied the same latent threshold spacings for giving partial credit in each of the four different test items.
4. Graded Response and Generalized Partial Credit models can be estimated by means of marginal or joint maximum likelihood as well as Bayesian estimation approaches, but not by conditional maximum likelihood.

C.12 Answers for Chapter 12

1. Advantages of computerized adaptive testing for the test takers are that the number of items that each test takers needs to answer is typically lower than in linear tests, and that the abilities of test takers can be estimated more precisely. This applies particularly to test takers with extremely high or low abilities if the item bank contains a sufficient number of suitable items.
2. Anchors in this context are items that are included in both of two test forms. They can be used as a means to place the item parameters from both forms on the same scale.
3. Two virtual items in the context of longitudinal IRT are actually the same item, presented at two different time points. Since changes in the person abilities will manifest as changes in the item parameters for different time points, the item parameters of the virtual items will reflect the item difficulty plus any changes over time.

References

Ackerman, T. (1992). A didactic explanation of item bias, item impact, and item validity from a multidimensional perspective. *Journal of Educational Measurement*, *29*(1), 67–91.

Adams, R., Wu, M., & Wilson, M. (2012). The Rasch rating model and the disordered threshold controversy. *Educational and Psychological Measurement*, *72*(4), 547–573.

Agresti, A. (2002). *Categorical data analysis.* Hoboken, NJ: John Wiley and Sons, Inc.

Albano, A. (2016). equate: An R package for observed-score linking and equating. *Journal of Statistical Software*, *74*(8), 1–36.

American Educational Research Association, American Psychological Association, & National Council on Measurement in Education. (2014). *Standards for educational and psychological testing.* Washington, DC: American Educational Research Association.

Andersen, E. (1973). A goodness of fit test for the Rasch model. *Psychometrika*, *25*(1), 123–140.

Andersson, B., Bränberg, K., & Wiberg, M. (2013). Performing the kernel method of test equating with the package kequate. *Journal of Statistical Software*, *55*(6), 1–25.

Andrich, D. (1978). A rating formulation for ordered response categories. *Psychometrika*, *3*(4), 357–374.

Andrich, D. (2004). Controversy and the Rasch model. *Medical Care*, *42*(1), I-7–I-16.

Andrich, D. (2010). Understanding the response structure and process in the polytomous Rasch model. In M. Nering & R. Ostini (Eds.), *Handbook of polytomous item response theory models.* New York, NY: Routledge.

Andrich, D., & Marais, I. (2019). *A course in Rasch measurement theory.* Singapore: Springer.

Baker, F., & Kim, S.-H. (2004). *Item response theory: Parameter estimation techniques* (2nd ed.). Boca Raton, FL: CRC Press.

Barton, M., & Lord, F. (1981). An upper asymptote for the three-parameter logistic item-response model. *ETS Research Report Series*, *1981*(1), i–8.

Bates, D., Mächler, M., Bolker, B., & Walker, S. (2015). Fitting linear mixed-effects models using lme4. *Journal of Statistical Software*, *67*(1), 1–48.

Bates, D., Mächler, M., Bolker, B., Walker, S., Christensen, R., Singmann, H., et al. (2022). *lme4: Linear mixed-effects models using 'Eigen' and*

S4. Retrieved from `https://CRAN.R-project.org/package=lme4` (R package).
Battauz, M. (2015). equateIRT: An R package for IRT test equating. *Journal of Statistical Software*, *68*(7), 1–22.
Battauz, M. (2021a). *equateIRT: An R package for IRT test equating.* Retrieved from `https://CRAN.R-project.org/package=equateIRT` (R package).
Battauz, M. (2021b). *equateMultiple: Equating of multiple forms.* Retrieved from `https://CRAN.R-project.org/package=equateMultiple` (R package).
Bechger, T., & Akkermans, W. (2001). A note on the equivalence of the graded response model and the sequential model. *Psychometrika*, *66*(3), 461–464.
Birnbaum, A. (1968). Some latent trait models and their use in inferring an examinee's ability. In F. Lord & M. Novick (Eds.), *Statistical theories of mental test scores.* Reading: Addison-Wesley.
Bock, D. (1972). Estimating item parameters and latent ability when responses are scored in two or more nominal categories. *Psychometrika*, *37*(1), 29–51.
Bond, T., & Fox, C. (2007). *Applying the Rasch model: Fundamental measurement in the human sciences.* Mahwah, NJ: Lawrence Erlbaum Assoc.
Brown, C., Templin, J., & Cohen, A. (2015). Comparing the two-and three-parameter logistic models via likelihood ratio tests: A commonly misunderstood problem. *Applied Psychological Measurement*, *39*(5), 335–348.
Bürkner, P.-C. (2017). brms: An R package for Bayesian multilevel models using Stan. *Journal of Statistical Software*, *80*(1), 1–28.
Bürkner, P.-C. (2021). Bayesian item response modeling in R with brms and Stan. *Journal of Statistical Software*, *100*(5), 1–54.
Cai, L. (2010). A two-tier full-information item factor analysis model with applications. *Psychometrika*, *75*(4), 581–612.
Camilli, G. (2006). Test fairness. In R. Brennan (Ed.), *Educational measurement* (pp. 221–256). ACE/Praeger Series on Higher Education.
Carroll, J. (1993). *Human cognitive abilities: A survey of factor-analytic results.* Cambridge, England: Cambridge University Press.
Casella, G., & Berger, R. (2002). *Statistical inference* (2nd ed.). Pacific Grove, CA: Duxbury.
Chalmers, R. P. (2016). Generating adaptive and non-adaptive test interfaces for multidimensional item response theory applications. *Journal of Statistical Software*, *71*(5), 1–39.
Chalmers, R. P. (2021). *mirt: Multidimensional item response theory.* Retrieved from `https://CRAN.R-project.org/package=mirt` (R package).
Chalmers, R. P., & Ng, V. (2017). Plausible-value imputation statistics for detecting item misfit. *Applied Psychological Measurement*, *41*(5), 372–387.

Chalmers, R. P., Pek, J., & Liu, Y. (2017). Profile-likelihood confidence intervals in item response theory models. *Multivariate Behavioral Research*, *52*(5), 533–550.

Chen, W.-H., & Thissen, D. (1997). Local dependence indexes for item pairs using item response theory. *Journal of Educational and Behavioral Statistics*, *22*(3), 265–289.

Cheng, Y., & Yuan, K. (2010). The impact of fallible item parameter estimates on latent trait recovery. *Psychometrika*, *75*(2), 280–291.

Christensen, K., Makransky, G., & Horton, M. (2017). Critical values for Yen's Q3: Identification of local dependence in the Rasch model using residual correlations. *Applied Psychological Measurement*, *41*(3), 178–194.

Cohen, A., & Bolt, D. (2005). A mixture model analysis of differential item functioning. *Journal of Educational Measurement*, *42*(2), 133–148.

Dayton, M., & Macready, G. (1988). Concomitant-variable latent-class models. *Journal of the American Statistical Association*, *83*(401), 173–178.

De Ayala, R. J. (2009). *The theory and practice of item response theory.* New York, NY: Guilford Press.

De Boeck, P. (2008). Random item IRT models. *Psychometrika*, *73*(4), 553–559.

De Boeck, P., Bakker, M., Zwitser, R., Nivard, M., Hofman, A., Tuerlinckx, F., et al. (2011). The estimation of item response models with the lmer function from the lme4 package in R. *Journal of Statistical Software*, *39*(12), 1–28.

De Boeck, P., & Wilson, M. (Eds.). (2004). *Explanatory item response models: A generalized linear and nonlinear approach.* New York, NY: Springer.

Debelak, R. (2019). An evaluation of overall goodness-of-fit tests for the Rasch model. *Frontiers in Psychology*, *9*, 2710.

Debelak, R., & Koller, I. (2020). Testing the local independence assumption of the Rasch model with Q3-based nonparametric model tests. *Applied Psychological Measurement*, *44*(2), 103–117.

DeMars, C. (2010). *Item response theory.* New York, NY: Oxford University Press.

Desjardins, C., & Bulut, O. (2018). *Handbook of educational measurement and psychometrics using R.* Boca Raton, FL: CRC Press.

Dorans, N., Pommerich, M., & Holland, P. (Eds.). (2007). *Linking and aligning scores and scales.* New York, NY: Springer.

Drasgow, F. (1989). An evaluation of marginal maximum likelihood estimation for the two-parameter logistic model. *Applied Psychological Measurement*, *13*(1), 77–90.

Drasgow, F., Levine, M., & Williams, E. (1985). Appropriateness measurement with polychotomous item response models and standardized indices. *British Journal of Mathematical and Statistical Psychology*, *38*(1), 67–86.

Draxler, C. (2010). Sample size determination for Rasch model tests. *Psychometrika, 75*(4), 708–724.

Draxler, C., & Alexandrowicz, R. (2015). Sample size determination within the scope of conditional maximum likelihood estimation with special focus on testing the Rasch model. *Psychometrika, 80*(4), 897–919.

Embretson, S., & Reise, S. (2000). *Item response theory for psychologists.* Mahwah, NJ: Lawrence Erlbaum Associates.

Engelhard, G. (2013). *Invariant measurement: Using Rasch models in the social, behavioral, and health sciences.* New York, NY: Routledge.

Finch, W. H., & French, B. F. (2015). *Latent variable modeling with R.* New York, NY: Routledge.

Fischer, G. H. (1973). The linear logistic test model as an instrument in educational research. *Acta Psychologica, 37*(6), 359–374.

Fischer, G. H. (1995). Linear logistic models for change. In G. H. Fischer & I. W. Molenaar (Eds.), *Rasch models: Foundations, recent developments and applications.* New York, NY: Springer.

Fischer, G. H., & Molenaar, I. W. (Eds.). (1995). *Rasch models: Foundations, recent developments and applications.* New York, NY: Springer.

Fischer, G. H., & Scheiblechner, H. (1970). Algorithms and programs for the probabilistic test model of Rasch. *Psychologische Beiträge, 12*(1), 23–51.

Formann, A. (1992). Linear logistic latent class analysis for polytomous data. *Journal of the American Statistical Association, 87*(418), 476–486.

Fox, J.-P. (2010). *Bayesian item response modeling: Theory and applications.* New York, NY: Springer Science & Business Media.

Frick, H., Strobl, C., Leisch, F., & Zeileis, A. (2012). Flexible Rasch mixture models with package psychomix. *Journal of Statistical Software, 48*(7), 1–25.

Frick, H., Strobl, C., & Zeileis, A. (2015). Rasch mixture models for DIF detection: A comparison of old and new score specifications. *Educational and Psychological Measurement, 75*(2), 208–234.

Gandrud, C. (2014). *Reproducible research with R and RStudio.* Boca Raton, FL: CRC Press.

Gelman, A., & Hill, J. (2006). *Data analysis using regression and multilevel/hierarchical models.* Cambridge: Cambridge University Press.

Gelman, A., Stern, J., Dunson, H., Vehtari, A., & Rubin, D. (2013). *Bayesian data analysis* (3rd ed.). Boca Raton, FL: CRC Press.

Glas, C. A. W., & Verhelst, N. (1995). Testing the Rasch model. In G. H. Fischer & I. W. Molenaar (Eds.), *Rasch models. Their foundations, recent developments and applications* (pp. 69–96). New York, NY: Springer.

Gustafsson, J. (1980). Testing and obtaining fit of data in the Rasch model. *British Journal of Mathematical and Statistical Psychology, 33*(2), 205–233.

Hartig, J., & Höhler, J. (2008). Representation of competencies in multidimensional IRT models with within- and between-item multidimensionality. *Zeitschrift für Psychologie*, *216*(2), 89–101.

Hoijtink, H., & Boomsma, A. (1995). On person parameter estimation in the dichotomous Rasch model. In G. H. Fischer & I. W. Molenaar (Eds.), *Rasch models. Their foundations, recent developments and applications* (pp. 53–68). New York, NY: Springer.

Holland, P., Thayer, D., Wainer, H., & Braun, H. (1988). Differential item performance and the Mantel-Haenszel procedure. In H. Wainer & H. Braun (Eds.), *Test validity* (pp. 129–145). Hillsdale, NJ: Lawrence Erlbaum.

Holling, H., Bertling, J., & Zeuch, N. (2009). Automatic item generation of probability word problems. *Studies in Educational Evaluation*, *35*(2–3), 71–76.

Hurvich, C., & Tsai, C.-L. (1993). A corrected Akaike information criterion for vector autoregressive model selection. *Journal of Time Series Analysis*, *14*(3), 271–279.

Husén, T. (Ed.). (1967). *International study of achievement in mathematics: A comparison of twelve countries* (Vols. 1–2). Stockholm: Almqvist & Wiksell.

Irtel, H. (1996). *Entscheidungs- und testtheoretische Grundlagen der psychologischen Diagnostik.* Bern, Switzerland: Peter Lang.

Kaplan, R., & Saccuzzo, D. (2013). *Psychological testing: Principles, applications and issues* (8th ed.). Independence, KY: Cengage Learning.

Keller, L., Swaminathan, H., & Sireci, S. (2003). Evaluating scoring procedures for context-dependent item sets. *Applied Measurement in Education*, *16*, 207–222.

Kiefer, T., Robitzsch, A., & Wu, M. (2013). *TAM tutorials: Home page.* Retrieved from `https://www.edmeasurementsurveys.com/TAM/Tutorials`.

Kolen, M., & Brennan, R. (Eds.). (2014). *Test equating, scaling, and linking: Methods and practices.* New York, NY: Springer.

Koller, I., Alexandrowicz, R., & Hatzinger, R. (2012). *Das Rasch Modell in der Praxis: Eine Einführung in eRm* (Vol. 3786). Wien: UTB.

Koller, I., & Hatzinger, R. (2013). Nonparametric tests for the Rasch model: Explanation, development, and application of quasi-exact tests for small samples. *Interstat*, *11*, 1–16.

Komboz, B., Strobl, C., & Zeileis, A. (2017). Tree-based global model tests for polytomous Rasch models. *Educational and Psychological Measurement*, *78*(1), 128–166.

Kopf, J., Zeileis, A., & Strobl, C. (2015a). Anchor selection strategies for DIF analysis: Review, assessment, and new approaches. *Educational and Psychological Measurement*, *75*(1), 22–56.

Kopf, J., Zeileis, A., & Strobl, C. (2015b). A framework for anchor methods and an iterative forward approach for DIF detection. *Applied Psychological Measurement*, *39*(2), 83–103.

Kruschke, J. (2014). *Doing Bayesian data analysis* (2nd ed.). Academic Press.

Linacre, J. (2002). What do infit and outfit, mean-square and standardized mean? *Rasch Measurement Transactions*, *16*(2), 878.

Linacre, J. (2020). *Winsteps.* Retrieved from `https://www.winsteps.com`.

Lunn, D., Spiegelhalter, D., Thomas, A., & Best, N. (2009). The BUGS project: Evolution, critique and future directions (with discussion). *Statistics in Medicine*, *28*, 3049–3082.

Lunn, D., Thomas, A., Best, N., & Spiegelhalter, D. (2000). WinBUGS – a Bayesian modeling framework: Concepts, structure and extensibility. *Statistics & Computing*, *10*, 325–337.

Magis, D. (2020). *difR: Collection of methods to detect dichotomous differential item functioning (DIF).* Retrieved from `https://CRAN.R-project.org/package=difR` (R package).

Magis, D., & Barrada, J. (2017). Computerized adaptive testing with R: Recent updates of the package catR. *Journal of Statistical Software, Code Snippets*, *76*(1), 1–19.

Magis, D., Béland, S., Tuerlinckx, F., & De Boeck, P. (2010). A general framework and an R package for the detection of dichotomous differential item functioning. *Behavior Research Methods*, *42*(3), 847–862.

Magis, D., & Raîche, G. (2012a). On the relationships between Jeffreys modal and weighted likelihood estimation of ability under logistic IRT models. *Psychometrika*, *77*(1), 163–169.

Magis, D., & Raîche, G. (2012b). Random generation of response patterns under computerized adaptive testing with the R package catR. *Journal of Statistical Software*, *48*(8), 1–31.

Magis, D., Yan, D., & von Davier, A. A. (2018). *mstR: Procedures to generate patterns under multistage testing.* Retrieved from `https://CRAN.R-project.org/package=mstR` (R package).

Mair, P. (2018). *Modern psychometrics with R.* Cham: Springer.

Mair, P., & Hatzinger, R. (2007). Extended Rasch modeling: The eRm package for the application of IRT models in R. *Journal of Statistical Software*, *20*(9), 1–20.

Mair, P., Hatzinger, R., & Maier, M. J. (2021). *eRm: Extended Rasch modeling.* Retrieved from `https://CRAN.R-project.org/package=eRm` (R package).

Masters, G. (1982). A Rasch model for partial credit scoring. *Psychometrika*, *47*(2), 149–174.

Maydeu-Olivares, A. (2013). Goodness-of-fit assessment of item response theory models. *Measurement: Interdisciplinary Research and Perspectives*, *11*(3), 71–101.

Maydeu-Olivares, A., Cai, L., & Hernández, A. (2011). Comparing the fit of item response theory and factor analysis models. *Structural Equation Modeling: A Multidisciplinary Journal*, *18*(3), 333–356.

Maydeu-Olivares, A., & Joe, H. (2006). Limited information goodness-of-fit testing in multidimensional contingency tables. *Psychometrika*, *71*(4), 713–732.

Maydeu-Olivares, A., & Montaño, R. (2013). How should we assess the fit of Rasch-type models? Approximating the power of goodness-of-fit statistics in categorical data analysis. *Psychometrika*, *78*(1), 116–133.

McDonald, R. (2011). *Test theory: A unified treatment.* New York, NY: Routledge.

McKinley, R., & Mills, C. (1985). A comparison of several goodness-of-fit statistics. *Applied Psychological Measurement*, *9*(1), 49–57.

Meiser, T. (2007). Rasch models for longitudinal data. In *Multivariate and mixture distribution Rasch models. Extensions and applications* (pp. 191–199). New York, NY: Springer.

Michell, J. (1999). *Measurement in psychology: A critical history of a methodological concept.* Cambridge University Press.

Molenaar, I. W. (1995). Estimation of item parameters. In G. H. Fischer & I. W. Molenaar (Eds.), *Rasch models. Their foundations, recent developments and applications* (pp. 39–51). New York, NY: Springer.

Muraki, E. (1992). A generalized partial credit model: Application of an EM algorithm. *Applied Psychological Measurement*, *16*(2), 159–176.

Orlando, M., & Thissen, D. (2000). Likelihood-based item-fit indices for dichotomous item response theory models. *Applied Psychological Measurement*, *24*(1), 50–64.

Orlando, M., & Thissen, D. (2003). Further investigation of the performance of S - X2: An item fit index for use with dichotomous item response theory models. *Applied Psychological Measurement*, *27*(4), 289–298.

Paek, I., & Cole, K. (2020). *Using R for item response theory model applications.* Abingdon, Oxon: Routledge.

Penfield, R. (2014). An NCME instructional module on polytomous item response theory models. *Educational Measurement: Issues and Practice*, *33*(1), 36–48.

Pinheiro, J., & Bates, D. (2000). *Mixed-effect models in S and S-PLUS.* Springer Science & Business Media.

Plummer, M. (2017). *JAGS: Just another Gibbs sampler.* Retrieved from `https://mcmc-jags.sourceforge.io`.

Ponocny, I. (2001). Nonparametric goodness-of-fit tests for the Rasch model. *Psychometrika*, *66*(3), 437–459.

R Core Team. (2021). *R: A language and environment for statistical computing.* Vienna, Austria. Retrieved from `https://www.R-project.org/` (Software Manual).

Rasch, G. (1960). *Probabilistic models for some intelligence and attainment tests.* Copenhagen: The Danish Institute of Educational Research.

Reckase, M. (2009). *Multidimensional item response theory.* New York, NY: Springer.

Rijmen, F., Tuerlinckx, F., De Boeck, P., & Kuppens, P. (2003). A nonlinear mixed model framework for item response theory. *Psychological Methods*, *5*(2), 185–205.

Rizopoulos, D. (2006). ltm: An R package for latent variable modelling and item response theory analyses. *Journal of Statistical Software*, *17*(5), 1–25.

Robitzsch, A., Kiefer, T., & Wu, M. (2021). *TAM: Test analysis modules.* Retrieved from `https://CRAN.R-project.org/package=TAM` (R package).

Rost, J. (1990). Rasch models in latent classes: An integration of two approaches to item analysis. *Applied Psychological Measurement*, *14*(3), 271–282.

Rost, J. (2004). *Testtheorie und Testkonstruktion.* Bern, Switzerland: Hans Huber.

Roussos, L., & Stout, W. (1996). A multidimensionality-based DIF analysis paradigm. *Applied Psychological Measurement*, *20*(4), 355–371.

Rulison, K., & Loken, E. (2009). I've fallen and I can't get up: Can high-ability students recover from early mistakes in CAT? *Applied Psychological Measurement*, *33*(2), 83–101.

Samejima, F. (1969). Estimation of latent ability using a response pattern of graded scores. *Psychometrika Monograph Supplement*, *34*(4).

Sclove, S. (1987). Application of model-selection criteria to some problems in multivariate analysis. *Psychometrika*, *52*(3), 333–343.

Sijtsma, K. (1998). Methodology review: Nonparametric IRT approaches to the analysis of dichotomous item scores. *Applied Psychological Measurement*, *22*(1), 3–31.

Sijtsma, K., & Hemker, B. (1998). Nonparametric polytomous IRT models for invariant item ordering, with respect to parametric models. *Psychometrika*, *63*(2), 183–200.

Sijtsma, K., & van der Ark, A. (2017). A tutorial on how to do a Mokken scale analysis on your test and questionnaire data. *British Journal of Mathematical and Statistical Psychology*, *70*(1), 137–158.

Sinharay, S. (2005). Assessing fit of unidimensional item response theory models using a Bayesian approach. *Journal of Educational Measurement*, *42*(4), 375–394.

Sinharay, S., Johnson, M., & Stern, H. (2006). Posterior predictive assessment of item response theory models. *Applied Psychological Measurement*, *30*(4), 298–321.

Spector, P. (2008). *Data manipulation with R.* New York, NY: Springer.

Stan Development Team. (2020). *rstanarm: Bayesian applied regression modeling via Stan.* Retrieved from `https://mc-stan.org/` (R package).

Stan Development Team. (2021). *RStan: The R interface to Stan.* Retrieved from `https://mc-stan.org/` (R package).

Stan Development Team. (2022). *Stan modeling language users guide and reference manual.* Retrieved from `https://mc-stan.org/` (Software Manual).

Steinfeld, J., & Robitzsch, A. (2021). Item parameter estimation in multistage designs: A comparison of different estimation approaches for the Rasch model. *Psych*, *3*(3), 279–307.

Stevens, S. (1946). On the theory of scales of measurement. *Science*, *103*, 677–680.

Stone, C. (1992). Recovery of marginal maximum likelihood estimates in the two-parameter logistic response model: An evaluation of multilog. *Applied Psychological Measurement*, *16*(1), 1–16.

Stone, C. (2000). Monte Carlo based null distribution for an alternative goodness-of-fit test statistic in IRT models. *Journal of Educational Measurement*, *37*(1), 58–75.

Strobl, C., Kopf, J., Kohler, L., von Oertzen, T., & Zeileis, A. (2021). Anchor point selection: Scale alignment based on an inequality criterion. *Applied Psychological Measurement*, *45*(3), 214–230.

Strobl, C., Kopf, J., & Zeileis, A. (2015). Rasch trees: A new method for detecting differential item functioning in the Rasch model. *Psychometrika*, *30*(2), 289–316.

Strobl, C., Malley, J., & Tutz, G. (2009). An introduction to recursive partitioning: Rationale, application and characteristics of classification and regression trees, bagging and random forests. *Psychological Methods*, *14*(4), 323–348.

Suárez-Falcón, J. C., & Glas, C. A. W. (2003). Evaluation of global testing procedures for item fit to the Rasch model. *British Journal of Mathematical and Statistical Psychology*, *56*(1), 127–143.

Sugiura, N. (1978). Further analysts of the data by Akaike's information criterion and the finite corrections. *Communications in Statistics - Theory and Methods*, *7*(1), 13–26.

Tennant, A., & Conaghan, P. G. (2007). The Rasch measurement model in rheumatology: What is it and why use it? When should it be applied, and what should one look for in a Rasch paper? *Arthritis Care & Research*, *57*(8), 1358–1362.

Thissen, D., & Steinberg, L. (1986). A taxonomy of item response models. *Psychometrika*, *51*(4), 567–577.

Thissen, D., Steinberg, L., & Wainer, H. (1988). Use of item response theory in the study of group difference in trace lines. In H. Wainer & H. I. Braun (Eds.), *Test validity* (pp. 147–170). Hillsdale, NJ: Lawrence Erlbaum.

Tsutakawa, R., & Johnson, J. (1990). The effect of uncertainty of item parameter estimation on ability estimates. *Psychometrika*, *55*(2), 371–390.

Tutz, G. (1990). Sequential item response models with an ordered response. *British Journal of Mathematical and Statistical Psychology*, *43*(1), 39–55.

Tutz, G. (1997). Sequential models for ordered responses. In W. van der Linden & R. K. Hambleton (Eds.), *Handbook of modem item response theory* (pp. 140–152). New York, NY: Springer.

van der Linden, W. (Ed.). (2016–2018). *Handbook of item response theory – Three volume set.* New York, NY: CRC Press.

van der Linden, W., & Glas, C. A. W. (Eds.). (2010). *Elements of adaptive testing.* New York, NY: Springer.

van der Linden, W., & Hambleton, R. (Eds.). (1997). *Handbook of modern item response theory.* New York, NY: Springer.

van Wyke, J., & Andrich, D. (2006). *A typology of polytomously scored mathematics items disclosed by the Rasch model: Implications for constructing a continuum of achievement* (Tech. Rep. No. 2). Australian Research Council Linkage Grant LP0454080.

von Davier, A. A. (Ed.). (2011a). *Statistical models for test equating, scaling, and linking.* New York, NY: Springer.

von Davier, A. A. (2011b). A statistical perspective on equating test scores. In A. A. von Davier (Ed.), *Statistical models for test equating, scaling, and linking* (pp. 1–17). New York, NY: Springer.

von Davier, M. (2020). TIMSS 2019 scaling methodology: Item response theory, population models, and linking across modes. In M. Martin, M. von Davier, & I. Mullis (Eds.), *Methods and procedures: TIMSS 2019 technical report.* Boston College, TIMSS & PIRLS International Study Center.

Wainer, H., Bradlow, E., & Wang, X. (2007). *Testlet response theory and its applications.* New York, NY: Cambridge University Press.

Wainer, H., Dorans, N., Flaugher, R., Mislevy, R., Thissen, D., Eignor, D., et al. (2000). *Computerized adaptive testing: A primer* (2nd ed.). New York, NY: Routledge.

Wang, W.-C., Shih, C.-L., & Sun, G.-W. (2012). The DIF-free-then-DIF strategy for the assessment of differential item functioning. *Educational and Psychological Measurement*, *72*(4), 687–708.

Warm, T. (1989). Weighted likelihood estimation of ability in item response theory. *Psychometrika*, *54*(3), 427–450.

Weijters, B., Baumgartner, H., & Schillewaert, N. (2013, 05). Reversed item bias: An integrative model. *Psychological Methods*, *18*.

Wetzel, E., & Carstensen, C. (2014). Reversed thresholds in partial credit models – A reason for collapsing categories? *Assessment*, *21*(6), 744–765.

Willse, J. (2014). *mixRasch: Mixture Rasch models with JMLE.* Retrieved from `https://CRAN.R-project.org/package=mixRasch` (R package).

Wright, B. (1977). Solving measurement problems with the Rasch model. *Journal of Educational Measurement*, *14*(2), 97–116.

Wright, B., & Masters, G. (1982). *Rating scale analysis.* Chicago, IL: MESA press.

Wright, B., & Masters, G. (1990). Computation of outfit and infit statistics. *Rasch Measurement Transactions*, *3*(4), 84–85.

Wright, B., & Stone, M. (1999). *Measurement essentials.* Wilmington: Wide Range Inc.

Yan, D., von Davier, A. A., & Lewis, C. (Eds.). (2016). *Computerized multistage testing: Theory and applications.* Princeton, NJ: CRC Press.

Yen, W. (1981). Using simulation results to choose a latent trait model. *Applied Psychological Measurement*, *5*(2), 245–262.

Yen, W. (1984). Effects of local item dependence on the fit and equating performance of the three-parameter logistic model. *Applied Psychological Measurement*, *8*(2), 125–145.

Yen, W. (1987). A comparison of the efficiency and accuracy of bilog and logist. *Psychometrika*, *52*(2), 275–291.

Zeileis, A., Strobl, C., Wickelmaier, F., Komboz, B., & Kopf, J. (2021). *psychotree: Recursive partitioning based on psychometric models.* Retrieved from `https://CRAN.R-project.org/package=psychotree` (R package).

Zeileis, A., Strobl, C., Wickelmaier, F., Komboz, B., Kopf, J., Schneider, L., et al. (2021). *psychotools: Psychometric modeling infrastructure.* Retrieved from `https://CRAN.R-project.org/package=psychotools` (R package).

Zwinderman, A., & van den Wollenberg, A. (1990). Robustness of marginal maximum likelihood estimation in the Rasch model. *Applied Psychological Measurement*, *14*(1), 73–81.

Author Index

Index